Edexcel GCSE (9-1)

Mathematics

Access to Foundation

Statistics and Geometry Workbook

ALWAYS LEARNING

PEARSON

Published by Pearson Education Limited, 80 Strand, London WC2R 0RL.

www.pearsonschoolsandfecolleges.co.uk

Text © Pearson Education Limited 2015
Typeset and illustrated by Tek-Art, West Sussex
Original illustrations © Pearson Education Limited 2015
Cover created by Fusako
Photography by NanaAkua

The right of Bobbie Johns, Peter Sherran and Glyn Payne to be identified as authors of this work has been asserted by them in accordance with the Copyright, Designs and Patents Act 1988.

First published 2015

17 16 15
10 9 8 7 6 5 4 3 2 1

British Library Cataloguing in Publication Data
A catalogue record for this book is available from the British Library

ISBN 978 1 447 99976 8

Printed in Slovakia by Neografia

Acknowledgements
Every effort has been made to contact copyright holders of material reproduced in this book. Any omissions will be rectified in subsequent printings if notice is given to the publishers.

A note from the publisher
In order to ensure that this resource offers high-quality support for the associated Pearson qualification, it has been through a review process by the awarding body. This process confirms that; this resource fully covers the teaching and learning content of the specification or part of a specification at which it is aimed. It also confirms that it demonstrates an appropriate balance between the development of subject skills, knowledge and understanding, in addition to preparation for assessment.

Endorsement does not cover any guidance on assessment activities or processes (e.g. practice questions or advice on how to answer assessment questions), included in the resource nor does it prescribe any particular approach to the teaching or delivery of a related course.

While the publishers have made every attempt to ensure that advice on the qualification and its assessment is accurate, the official specification and associated assessment guidance materials are the only authoritative source of information and should always be referred to for definitive guidance.

Pearson examiners have not contributed to any sections in this resource relevant to examination papers for which they have responsibility.

Examiners will not use endorsed resources as a source of material for any assessment set by Pearson.

Endorsement of a resource does not mean that the resource is required to achieve this Pearson qualification, nor does it mean that it is the only suitable material available to support the qualification, and any resource lists produced by the awarding body shall include this and other appropriate resources.

Notices

The calculator symbol shows questions where:
- calculator skills are being developed, or
- using a calculator effectively is an important aspect of answering the question, or
- the calculation exceeds the scope of written methods in earlier chapters.

Contents

Self-assessment chart   iv

Introduction   v

Unit 1   Data   1
   1.1   Types of data   1
   1.2   Data collection   1

Unit 2   Displaying data   6
   2.1   Pictograms   6
   2.2   Bar charts   8
   2.3   Line graphs   10
   2.4   Dual bar charts   11
   2.5   Two-way tables   13
   2.6   Pie charts   15

Unit 3   Calculating with data   19
   3.1   Averages and ranges   19

Unit 4   Interpreting data   25
   4.1   Read and interpret data presented in tables   25
   4.2   Interpret charts and graphs   29
   4.3   Find totals and modes from frequency tables or diagrams   36

Unit 5   Probability   41
   5.1   Use and interpret a probability scale   41
   5.2   Write down theoretical and experimental probabilities   43
   5.3   List outcomes   46

Unit 6   Measures   49
   6.1   Metric measures   49
   6.2   Convert between metric units   50
   6.3   Add and subtract units of measure   51
   6.4   Read scales   53
   6.5   Draw and measure lines   55

Unit 7   Angles   58
   7.1   Angles and turning   58
   7.2   Naming, measuring and drawing angles   59

Unit 8   Angle calculations   65
   8.1   Angles on a straight line and angles round a point   65
   8.2   Triangle properties   67

Unit 9   Constructing triangles   71
   9.1   Accurate drawings   71

Unit 10   Transformations   78
   10.1   Reflection   78
   10.2   Enlargement   79
   10.3   Congruent shapes and similar shapes   81

Unit 11   Circle and quadrilateral definitions   85
   11.1   Circles   85
   11.2   Quadrilaterals   88

Unit 12   Perimeter, area and volume   91
   12.1   Perimeter of rectangles   91
   12.2   Area of rectangles   93
   12.3   Volume of cuboids   94

Statistics Test   97

Geometry Test   104

Self-assessment chart

	Need more practice	Almost there	Got it!	
Unit 1 Data				**Step**
1.1 Types of data	☐	☐	☐	➡ 4th
1.2 Data collection	☐	☐	☐	➡ 4th
Unit 2 Displaying data				
2.1 Pictograms	☐	☐	☐	➡ 1st
2.2 Bar charts	☐	☐	☐	➡ 1st
2.3 Line graphs	☐	☐	☐	➡ 3rd
2.4 Dual bar charts	☐	☐	☐	➡ 3rd
2.5 Two-way tables	☐	☐	☐	➡ 4th
2.6 Pie charts	☐	☐	☐	➡ 4th
Unit 3 Calculating with data				
3.1 Averages and ranges	☐	☐	☐	➡ 3rd
Unit 4 Interpreting data				
4.1 Read and interpret data presented in tables	☐	☐	☐	➡ 4th
4.2 Interpret charts and graphs	☐	☐	☐	➡ 4th
4.3 Find totals and modes from frequency tables or diagrams	☐	☐	☐	➡ 4th
Unit 5 Probability				
5.1 Use and interpret a probability scale	☐	☐	☐	➡ 3rd
5.2 Write down theoretical and experimental probabilities	☐	☐	☐	➡ 4th
5.3 List outcomes	☐	☐	☐	➡ 4th
Unit 6 Measures				
6.1 Metric measures	☐	☐	☐	➡ 2nd
6.2 Convert between metric units	☐	☐	☐	➡ 3rd
6.3 Add and subtract units of measure	☐	☐	☐	➡ 3rd
6.4 Read scales	☐	☐	☐	➡ 3rd
6.5 Draw and measure lines	☐	☐	☐	➡ 1st
Unit 7 Angles				
7.1 Angles and turning	☐	☐	☐	➡ 1st
7.2 Naming, measuring and drawing angles	☐	☐	☐	➡ 3rd
Unit 8 Angle calculations				
8.1 Angles on a straight line and angles round a point	☐	☐	☐	➡ 4th
8.2 Triangle properties	☐	☐	☐	➡ 3rd
Unit 9 Constructing triangles				
9.1 Accurate drawings	☐	☐	☐	➡ 4th
Unit 10 Transformations				
10.1 Reflection	☐	☐	☐	➡ 3rd
10.2 Enlargement	☐	☐	☐	➡ 3rd
10.3 Congruent shapes and similar shapes	☐	☐	☐	➡ 3rd
Unit 11 Circle and quadrilateral definitions				
11.1 Circles	☐	☐	☐	➡ 1st
11.2 Quadrilaterals	☐	☐	☐	➡ 4th
Unit 12 Perimeter, area and volume				
12.1 Perimeter of rectangles	☐	☐	☐	➡ 4th
12.2 Area of rectangles	☐	☐	☐	➡ 4th
12.3 Volume of cuboids	☐	☐	☐	➡ 4th

Introduction

Helping you prepare for *Edexcel GCSE (9-1) Mathematics – Foundation*, this workbook is a good way to refresh your learning on Statistics and Geometry.

Work your way through this book unit by unit:

* The clear **learning objectives** help you focus
* The **key points** give you reminders
* The **worked examples** guide you through to the solution
* All the carefully stepped **practice** develops your confidence
* Stretch yourself a bit with **extend** questions
* The unit **summaries** help you recap and revise
* Take the **unit tests** to check your fluency and build your confidence
* Take the **Statistics and Geometry Tests** at the end of the book to check your progress.

And there's a useful self-assessment chart on page iv for you to fill in as you go!

Number

As further preparation alongside Geometry and Statistics, before you progress on to the Foundation GCSE course, you could also refresh your learning in Number and a little Algebra.

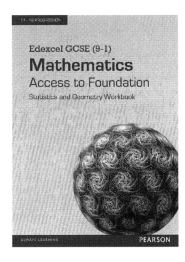

Types of data

1.1

By the end of this section you will know how to:
* Recognise and describe different types of data

Key points

* **Discrete data** can only take specific values.
* **Continuous data** can take any numerical value in a given range.

Example

1 Match each example of data with its type.

The numbers thrown on a dice ⟶ Discrete

The time taken to run 100 metres

The colour of a car

The weight of an apple ⟶ Continuous

> **Hint**
> Distance, weight, time and temperature are all continuous.

Practice

2 Match each example of data with its type.

  a The number of sweets in a jar

  b The temperature inside an oven

  c The wingspan of a butterfly   Discrete

  d The number of pages in a book

  e The number of students in a class   Continuous

  f The height of flowers in a garden

Data collection

1.2

By the end of this section you will know how to:
* Design and use a data collection sheet for discrete or continuous data

Key points

* A simple **data collection sheet** has a **data column**, a **tally column** and a **frequency column**. The frequency column is where you put the **totals** of the tally marks.
* **Discrete data** may be shown as a list of values in the data column.
* **Discrete data** may be grouped into class intervals in the data column.
* **Continuous data** is always grouped into **class intervals** in the data column. A class interval contains the numbers between its end-points.

1 A dice is rolled 60 times. The numbers rolled are shown below.

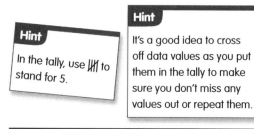

6	5	5	4	1	3	6	2	2	2	5	6
3	1	1	3	5	4	5	2	3	4	1	6
2	5	3	6	4	4	5	5	5	6	1	2
3	3	2	1	6	5	6	3	2	4	1	1

a Complete the table to show this data.

Number rolled	Tally	Frequency
1	||	
2	|	
3	|||	
4	|||	
5	|	
6	||	

> **Hint**
> In the tally, use ⊥⊥⊥ to stand for 5.

> **Hint**
> It's a good idea to cross off data values as you put them in the tally to make sure you don't miss any values out or repeat them.

> **Hint**
> Once the tally marks have been entered, count them to give the frequency for each number rolled.

b Find the total of the numbers in the **Frequency** column.

Total = ...

c Explain how your answer to part **b** may be used to check your answer to part **a**.

....

....

2 At a charity event, people were asked to guess the number of balloons inside a car. The results are shown below.

387	410	306	425	450	386	390	421	480	395
367	409	465	382	340	375	400	420	478	365
391	372	408	450	432	387	324	305	376	428
387	361	472	436	424	381	379	401	432	360
427	348	375	429	460	384				

a Complete the table to show this data.

Guess	Tally	Frequency
301–330	|	
331–360		
361–390	|||	
391–420	||	
421–450	|||	
451–480	|	

> **Hint**
> There are too many different values to list them all in the table. It is better to group the values into class intervals instead.

> **Hint**
> 301–330 is one class interval. This data is grouped into six class intervals.

b How many people tried to guess the number of balloons?

....

3 The heights of 30 students in Year 11 were measured.
The results (in metres) are shown below.

~~1.78~~ ~~1.69~~ ~~1.83~~ ~~1.74~~ ~~1.85~~ ~~1.93~~ ~~1.76~~ ~~1.65~~ ~~1.69~~ ~~1.74~~
1.82   1.91   1.87   1.72   1.92   1.86   1.75   1.63   1.70   1.86
1.65   1.76   1.82   1.89   1.78   1.72   1.84   1.79   1.68   1.82

Complete the table to show this data.

Height (h m)	Tally	Frequency
$1.60 < h \leq 1.65$		
$1.65 < h \leq 1.70$	||	
$1.70 < h \leq 1.75$	||	
$1.75 < h \leq$	||	
$1.80 < h \leq 1.85$	||	
.... $< h \leq 1.90$	|	
.... $< h \leq 1.95$	|	

> **Hint**
> The class $1.65 < h \leq 1.70$ includes 1.70 but not 1.65.

4 This spinner is spun 40 times. The results are shown below.

4   3   3   5   2   4   1   2   5   1
5   3   2   2   1   4   5   1   4   3
5   1   1   2   3   2   4   2   5   4
4   2   2   4   2   1   5   3   3   5

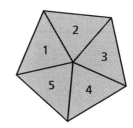

a Complete this table for the spinner.

Number spun	Tally	Frequency

b Which number has the highest frequency?   ....

5 30 students were asked to keep track of the number of text messages they sent in one week.
The results are shown below.

72   91   45   136   28   15   184   102   123   87
36   175   64   83   89   72   134   128   75   192
88   46   22   16   165   34   173   109   86   21

a Complete this table for the numbers of texts sent.

Number of texts	Tally	Frequency
11–40		
41–80		
81–110		
111–150		
151–200		

> **Hint**
> The classes do not have to be the same width.

b Explain why it is sensible to group the data in this case.

....

....

....

c How many students sent more than 150 texts? ....

6 The numbers of goals scored in a sample of 40 Premier League football games are shown below.

2	0	1	0	4	2	0	3	1	2
1	1	2	1	3	2	5	1	2	3
2	0	4	2	1	1	4	3	2	0
6	3	0	0	2	1	2	4	1	2

Design and complete a data collection sheet for this data.

Don't forget!

* Discrete data can only take .... values.

* Continuous data can take any .... value in a given ....

* On a data collection sheet, continuous data is always .... into .... intervals.

Unit test

1 Suki lives near a main road. She decides to do a survey of the first 50 vehicles that pass by her house. Her results are shown below.

Car	Car	Truck	Car	Car	Truck	Van	Van	Car	Truck
Car	Van	Car	Car	Bus	Car	Van	Car	Car	Van
Car	M/bike	M/bike	Car	Van	Truck	Truck	Van	Car	Car
Van	Car	Car	Truck	Car	Car	Truck	Truck	Van	Car
Car	Bus	Truck	Car	Car	Car	Car	Car	Car	Van

a Design and complete a data collection sheet for Suki's data.

b What kind of data has Suki collected? ....

2 A biology student measures the lengths, in cm, of some leaves from an oak tree. Her data is shown below.

14.2	18.6	13.3	17.5	12.8	16.4	18.1	19.6	17.8	16.4
12.9	19.7	17.6	19.3	14.8	16.5	14.9	15.3	12.8	18.2
15.7	18.3	19.4	18.7	17.9	14.5	16.2	13.8	19.8	16.6

> **Hint**
> Start with the class interval $12.0 < l \leqslant 13.0$, where l represents length.

Design and complete a data collection sheet for this data.

Pictograms

2.1

By the end of this section you will know how to:

* Draw pictograms

Key points

* A **pictogram** uses symbols or pictures to represent numbers of items or events.
* A pictogram should have a **title**.
* A pictogram should have a **key** showing the value of each symbol.
* **Fractions** of a symbol may be used for smaller amounts. Choose a symbol that is easily divided into smaller parts that can be understood.

1 The table shows the numbers of people attending a school production during one week. Each figure has been rounded to the nearest 10.

Day	Number of people
Monday	120
Tuesday	90
Wednesday	100
Thursday	80
Friday	130
Saturday	150

Complete the pictogram to show this data.

Pictogram showing people attending the school production

Monday	☺☺☺☺☺☺☺☺☺☺☺☺
Tuesday	☺☺☺☺☺☺☺☺☺
Wednesday	
Thursday	
Friday	
Saturday	

Key: ☺ represents 10 people

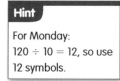

Hint

For Monday:
$120 \div 10 = 12$, so use 12 symbols.

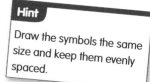

Hint

Draw the symbols the same size and keep them evenly spaced.

2 Show how to represent the following numbers using the key shown.

Key: ⊞ represents 4 boxes

a 6 boxes   ⊞⊟

b 7 boxes   ....

c 10 boxes   ....

d 13 boxes   ....

Practice

3 Jamie owns a busy restaurant.
He records the number of customers at different times on a Saturday night.
The table shows his results.

Time	Number of customers
7 pm	24
8 pm	38
9 pm	46
10 pm	32
11 pm	27

Hint

Calculate the number of symbols needed for each time before you draw the pictogram.

Show Jamie's data on a pictogram using the key shown.

Key: ⊕ represents 4 customers

Extend

4 The table shows the number of letters received by a company each day for a week.

Day	Number of letters
Monday	17
Tuesday	12
Wednesday	15
Thursday	11
Friday	14

Using the data shown in the pictogram for Monday, work out and complete the key.
Complete the pictogram.

Day	Number of letters received
Monday	☐ ☐ ☐ ☐ ☐ ☐ ☐ ☐ ☐
Tuesday	
Wednesday	
Thursday	
Friday	

Hint

8.5 rectangles are used here for 17 letters.

Key: ☐ represents .... letters

Bar charts

2.2

By the end of this section you will know how to:

✷ Draw bar charts

Key points

✷ The **height** or **length** of each bar shows the **value represented**.

✷ The bars may be horizontal or vertical.

✷ There should be a gap between the bars.

✷ The bars should all be the same width.

✷ The axes on a bar chart should be clearly labelled.

Example

1 Jess has a jar of sweets with coloured wrappers.
She counts the number of each colour and draws a frequency table.

Colour	Red	Blue	Green	Orange	Purple
Frequency	7	8	5	3	9

Complete the bar chart using the data and write a title.

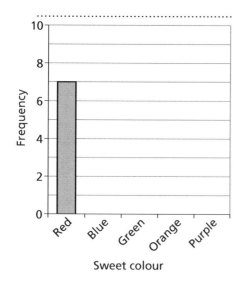

> **Hint**
>
> The bars should have the same width. Leave a gap between them.

2 Fill in the missing values in these tables.

a Jimmy has 34 marbles of different colours as shown in the table.

Colour	Blue	Green	Yellow	Red
Frequency	8	3	9	....

> **Hint**
>
> When you add the frequency values the total must match the total number of marbles (34).

b The table shows how Jo spent her wages.

Item	Rent	Food	Clothes	Entertainment	Other
Percentage	28	24	16	21	....

> **Hint**
>
> The percentages in the table add up to 100.

Practice

3 Forty people were asked about their favourite type of television programme.

Programme	Comedy	Drama	Soap	Film	Documentary
Frequency	11	6	10	....	4

Complete the table and bar chart.

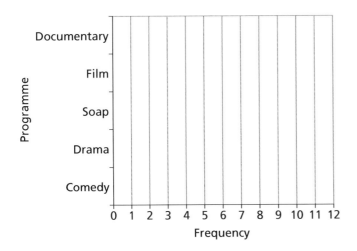

Extend

4 The pictogram shows how many coins of each type John has saved.

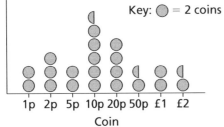

Draw a bar chart to represent the same information.

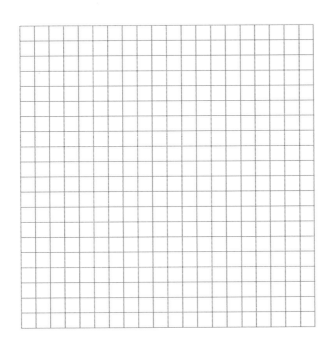

Line graphs

2.3

By the end of this section you will know how to:

* Draw line graphs

Key points

* A **line graph** is used to display pairs of data values by plotting them as coordinates.

* Points between the plotted points may have no meaning.

* The plotted points are joined with straight lines.

Example

1 Kate recorded the number of hours of sunshine each day of her holiday. Her results are shown in the table.

Day	Sat	Sun	Mon	Tue	Wed	Thu	Fri
Hours of sunshine	4	1	6	8	5	4	7

On the grid, draw a line graph for this information.

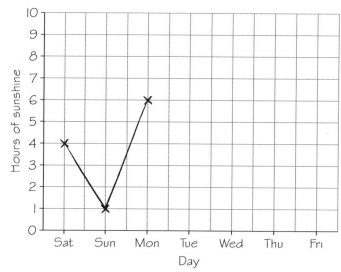

Hint

The first three points have been plotted for you. Each point is joined to the next with a straight line.

Practice

2 A campsite has tents of different sizes for different numbers of people. The table shows how many tents of different sizes are used one night.

Tent size (number of people in tent)	1	2	3	4	5
Frequency	5	12	15	20	16

Draw a line graph for this data.

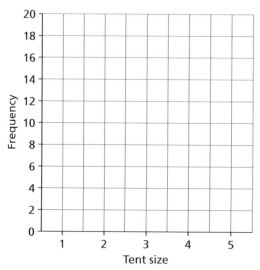

3 Jim checked his pulse rate, in beats per minute (bpm), every 10 minutes during a workout. The table shows his results.

Time (minutes)	Pulse rate (bpm)
0	68
10	157
20	126
30	110
40	145
50	128
60	96

On the grid, draw a line graph to show the information in the table.

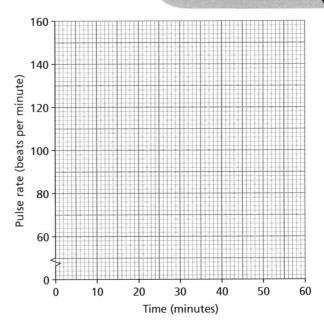

Dual bar charts

2.4

By the end of this section you will know how to:

✳ Draw dual bar charts

Key points

✳ The bars are **grouped** next to each other in every category.

✳ The grouping of the bars allows **comparisons** to be made within a category.

✳ There is a gap between one group of bars and the next.

✳ The bars may be **horizontal** or **vertical**.

✳ A dual bar chart should have a key explaining the different bars.

1 In the run-up to the end-of-course exams a teacher gives his class five tests. For each one, he records the highest and lowest scores for the class. The table shows the results.

Test	1	2	3	4	5
Highest	73	77	76	82	88
Lowest	36	45	40	53	64

Draw a dual bar chart to show the test results.

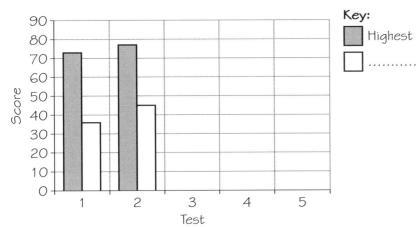

Key:
▨ Highest
☐ ....

2 A supermarket sells tubs of ice cream in two flavours, vanilla and chocolate.
The table shows how many of each flavour were sold in one week.

Day	Mon	Tue	Wed	Thu	Fri
Vanilla	12	14	10	9	18
Chocolate	11	12	14	7	21

Draw a vertical dual bar chart to show this data.

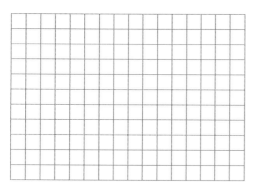

3 A mixed group of boys and girls were asked to name their favourite sports.
The results are shown in the table.

Sport	Football	Hockey	Tennis	Swimming	Athletics
Boys	10	1	1	2	3
Girls	4	4	3	5	3

Draw a horizontal dual bar chart to show these results.

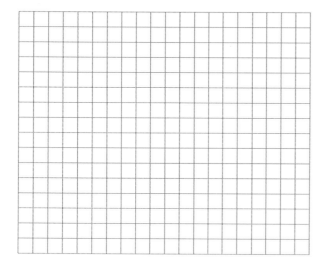

4 A theme park has rides for adults and children. The table shows the numbers of adult and child tickets sold during one week to the nearest 100.

Day	Mon	Tue	Wed	Thu	Fri	Sat	Sun
Adult	700	200	400	1600	2500	3000	2000
Child	1000	500	800	2000	3000	3500	2700

Show this information on a dual bar chart.
Use the grid on the opposite page.

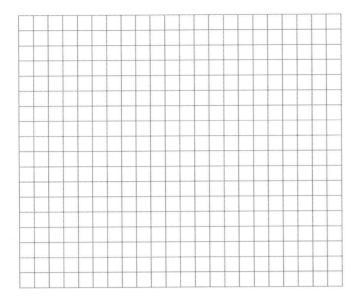

Need more practice ☐   Almost there ☐   Got it! ☐

Two-way tables

2.5

By the end of this section you will know how to:

✶ Draw two-way tables

Key points

✶ A **two-way table** has **labelled rows** and **columns**.

✶ Every number is in a row and a column so it shows **two types of information**.

✶ The total of all the rows is equal to the total of all the columns.

Example

1 Some Year 11 students were asked which subjects they would like to study in the sixth form. Their responses are shown below.

Boys

Maths	Phys	Chem	Biol	Biol	Maths	Eng	Hist	Geog	A̶r̶t̶	Econ
Geog	Maths	Econ	Phys	Eng	Maths	Biol	Geog	Hist	Hist	Eng

Girls

Eng	Biol	A̶r̶t̶	Maths	Chem	Eng	Biol	Maths	Hist	Geog
Maths	Biol	Econ	Hist	A̶r̶t̶	A̶r̶t̶	Maths	Hist	Eng	Biol

Show this information in a two-way table.

	Art	Biol	Chem	Econ	Eng	Geog	Hist	Maths	Phys	Total
Boys	1									
Girls	3									
Total	4									

Hint

Deal with one subject at a time. Check that the total in the bottom right-hand corner is the same vertically and horizontally.

2 This two-way table shows some information about the languages students take in the sixth form.

	French	German	Spanish	Total
Boys	7	6	....	18
Girls	8	....	9	24
Total	15	....	....	....

Hint

Use the totals in the rows and columns to find the missing values.

Complete the table.

3 Carol did a survey of favourite pets.
Here are her results.

Boys

dog	dog	cat	fish	rat	dog	rabbit
mouse	hamster	dog	cat	rat	fish	mouse
cat	dog	hamster	dog	fish	rabbit	dog

Girls

cat	mouse	dog	dog	mouse	rabbit	hamster
dog	rabbit	cat	mouse	rabbit	hamster	dog
cat	hamster	dog	mouse	fish	cat	rabbit

Show Carol's results in a two-way table.

4 In a survey, some Year 11 students were asked what they intended to do the following year. The two-way table below shows some information about their answers.

	Sixth form	College	Work	Total
Boys	24		11	58
Girls		29		
Total	52		27	

Complete the two-way table.

5 A maths teacher constructed the following two-way table based on 31 students in her class.

	Did homework	Did not do homework	Total
Did bring equipment	24		
Did not bring equipment			5
Total	25		

Complete the two-way table.

Practice

Extend

2.6 Pie charts

By the end of this section you will know how to:

✱ Draw pie charts

Key points

* A **pie chart** is a **circle** divided into **sectors**.
* Each sector represents a **category** of data.
* The **angle** of a sector represents the **size of the category** compared to the whole.
* A pie chart is useful for **comparing** a small number of categories.

Example

1 The table shows how a class of 30 students travel to school.

Method of travel	Number of students	Angle of pie chart
Bus	14	12° × 14 = 168°
Car	5	12° × 5 =
Cycle	3	
Walk	8	
Total	30	

360° ÷ 30 = 12°

Hint

Calculate the number of degrees that represents each student. Remember there are 360° in a circle.

Complete the table and pie chart.

Hint

Check that all the angles add up to 360°.

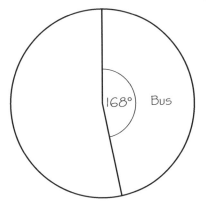

Hint

Use a protractor to measure the angles carefully at the centre of the circle.

Hint

Make sure that you have all the equipment you need for the examination.

Practice

2 A survey of the types of fuel used in 120 cars produced the results shown in the table.

Fuel type	Number of vehicles	Angle of pie chart
Petrol	72	
Diesel	31	
Dual fuel	2	
Hybrid	15	
Total		

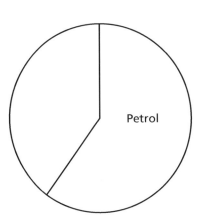

Complete the table and pie chart to show this information.

3 A waste recycling plant separates different types of waste into containers. The table shows the number of containers needed for each type in one month.

Waste type	Number of containers	Angle of pie chart
General waste	30	
Metal	8	
Organic	16	
Paper	6	
Wood	12	
Total		

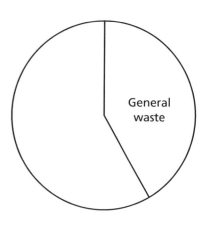

Complete the table and pie chart to show this information.

Extend

4 The table below shows how one student spent her day.

Activity	Number of hours	Angle of pie chart
Sleeping	9	
Working	7	
Dining	2	
Relaxing	6	
Total		

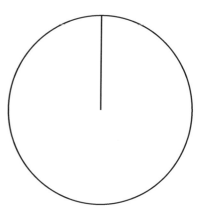

Complete the table and pie chart to show this information.

Don't forget!

✳ A pictogram should have a .... showing the value of the symbols.

✳ Bar charts must have a .... between the bars.

✳ A line graph may be used to display ungrouped .... data.

✳ A dual bar chart allows .... to be made within a category.

✳ In a two-way table, the .... of the rows = the total of the ....

✳ A pie chart is a useful way to compare a .... number of categories.

Unit test

1   The pictogram shows the number of hours of sunshine on five days during one week.

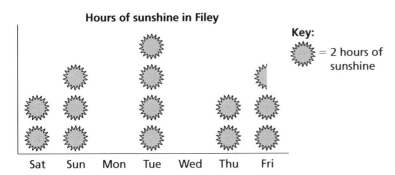

Hours of sunshine in Filey

Key:
= 2 hours of sunshine

Sat   Sun   Mon   Tue   Wed   Thu   Fri

a   Complete the pictogram using the information that there were two hours of sunshine on Monday and six hours on Wednesday.

b   Draw a bar chart using the grid below to show the same information.

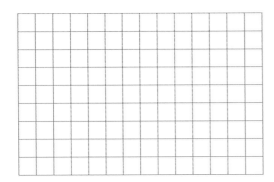

2   115 people responded to a survey about membership of a new gym.
The two-way table shows some information about their answers.

	Junior	Adult	Senior	Total
Standard	36	16		57
Full	5	18	10	
Premium		25		
Total	41			

Complete the two-way table.

3 Ade, Baz, Col, Dan and Eve ran two laps of a cross-country course.
   The table shows the time in minutes that they each took for each lap.

Name	Ade	Baz	Col	Dan	Eve
Lap 1	38	43	41	37	45
Lap 2	34	52	54	32	45

Draw a dual bar chart to show this data.

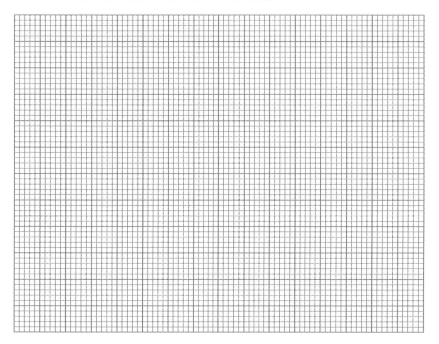

4 Sharon has 45 books. The table shows the categories that they fit into.

Book category	Number of books	Angle of pie chart
Comedy	12	
Thriller	7	
Mystery	9	
Romance	17	

In the circle, draw a pie chart to represent this information.

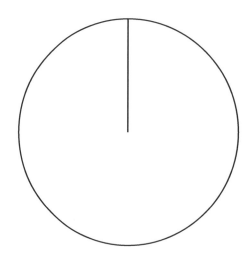

Averages and ranges

3.1

By the end of this section you will know how to:

✳ Find the mode, median, mean and range of a set of data

Key points

✳ There are three types of average: mode, median and mean.

✳ The **mode** is the value that occurs **most often** in the data.

✳ The **median** is the **middle value** once the data is put in order of size.

✳ If there are two middle values, the median is **halfway** between them.

✳ The **mean** is found by **adding** all of the data values together and **dividing** by the number of them.

✳ The **range** measures the **spread** of data. It is the difference between the largest and smallest values in the data.

Example

1 Helen wanted to know about the colours used for new cars. She looked at ten new cars in a car park and recorded these colours.

| white | blue | white | red | green |
| black | white | red | black | blue |

Find the mode.

> **Hint**
> The mode is the only average that does not have to be a number.

> **Hint**
> Which colour appears more than any other in the list?

Mode = ...

2 Paul counts the number of people in each car that passes. Here are his results.

| 2 | 1 | 1 | 3 | 2 | 1 |
| 1 | 2 | 3 | 1 | 2 | 1 |

Find the mode of Paul's data.

> **Hint**
> Look for the number that occurs most often.

Mode = ...

Practice

3 Kath spun a five-sided spinner 10 times. Here are her results.

3   5   4   4   3   1   4   2   5   2

Write down the mode.

Mode = ...

4 Adele counted the number of people at a bus stop at different times of the day. Here are her results.

1   0   3   4   3   2   0   3   5   2   3   1

Write down the mode.

Mode = ...

Extend

5 The numbers of matches in a sample of ten boxes are shown below.

49   48   46   50   48   47   50   48   49   46

Find the mode.

Mode = ...

6 Lizzy asked ten pupils to name their favourite colour. Here are her results.

green   black   red   blue   brown   blue   red   blue   black   white

Write down the mode.

Mode = ...

Example

7 Find the median of these nine numbers.

> 12   18   11   9   16   17   15   16   18

Write the numbers in order: 9   11   12   ....

The median is ....

> **Hint**
> The median is the middle value.

8 Find the median of these eight numbers.

> 8   9   7   10   7   9   8   10

Write the numbers in order: ....

The numbers in the middle are .... and ....

The median is ....

> **Hint**
> The number of values is even, so there is no middle value.

> **Hint**
> The median is halfway between the pair of numbers in the middle.

Practice

9 Here are the heights of the members of the Year 11 football team in centimetres.

> 183   176   175   180   182   180   181   172   178   185   179

Find the median height.

Median height = ... cm

10 Ten children were asked how many Easter eggs they had received. Here are the results.

> 11   7   6   7   10   9   6   9   7   12

Find the median.

Median number of eggs = ...

Extend

11 Eight Year 11 boys tried to do as many press-ups as possible in one attempt. Here are the results.

> 23   16   14   28   11   36   57   19

Find the median.

Median number of press-ups = ...

12 The cooking times in minutes of ten frozen ready meals are shown below.

> 11   16   18   12   17   20   15   10   15   16

Find the median.

Median cooking time = ... minutes

Example

13 Work out the mean of these eight football scores.

3   0   2   1   5   2   1   2

Total = 3 + 0 + 2 + ... = ...

Mean = ... ÷ .... = ...

> **Hint**
> Divide the total by the number of scores.

14 Jackie went ten-pin bowling. Her first six attempts produced these scores.

3   5   1   7   4   10

Calculate Jackie's mean score.

Total = ...

Mean = ... ÷ .... = ...

Practice

15 Here are the numbers of letters that Tom received during a six-day period.

4   3   6   5   4   8

Calculate the mean number of letters Tom received each day.

Mean = ...

16 Leslie made nine bunches of flowers.
Here are the numbers of flowers she used in the bunches.

10   12   11   15   13   16   12   8   11

Work out the mean number of flowers per bunch.

Mean = ...

Extend

17 Eight swimmers took part in a sponsored swim.
The numbers of lengths they completed are shown below.

21   16   25   30   34   18   22   26

Calculate the mean number of lengths each person swam.

Mean = ...

18 Here are the numbers of people staying in ten caravans.

4   3   2   5   4   4   3   2   6   3

Work out the mean number of people per caravan.

Mean = ...

19 The monthly salaries of eight employees are shown below.

£965   £824   £1026   £872   £1105   £988   £1009   £978

Work out the range.

Largest salary = ...   Smallest salary = ...

Range = ...   −   ....   = ...

> **Hint**
> The range is the difference between the largest salary and the smallest salary.

20 Five friends each threw one dart at a dart board. Their scores are shown below.

19   1   20   60   18

Work out the range.

Range = ...

21 Seven pupils gave these estimates of the height of a building in metres.

10   7   8.6   10.5   9   9.4   8

Work out the range.

Range = ...

22 Six members of a running club ran a half marathon. Here are their times.

1 h 26 min   1 h 15 min   2 h 11 min   1 h 45 min   1 h 52 min   1 h 38 min

Work out the range.

Range = ...

23 In a strongman contest, competitors had to hurl barrels over a steel wall. The times taken were:

1 min 18 sec   1 min 12 sec   23 sec   1 min 17 sec   21 sec   1 min 11 sec

Work out the range.

Range = ...

24 The lowest overnight temperatures in eight UK cities in January were:

−2°C   0°C   −4°C   −7°C   −3°C   −8°C   −6°C   −12°C

Work out the range.

Range = ...

Don't forget!

✳ The mode, median and mean are three types of ....

✳ The mode is the value that occurs ....

✳ To find the median, first put the values in ....

✳ To find the mean, divide the .... by the

....

✳ The .... measures the spread of the data.

Unit test

1 Eleven adults were asked how many close friends they had. Here are the results.

1   3   1   2   4   2   1   3   1   2   5

a Find the mode.

Mode = ...

b Find the median.

Median = ...

c Work out the mean.

Mean = ...

2 In a gym test, the 11 members of a football team did as many sit-ups as they could in one minute. Here are the results.

48   72   55   64   58   52   49   56   70   64   61

a Work out the mean.

Mean = ...

b Find the median.

Median = ...

c Work out the range.

Range = ...

3 Bill recorded the times taken (in seconds) by a sample of ten cars to travel along a stretch of motorway. Here are his results.

26   23   22   19   23   21   23   22   25   26

a Find the mode.

Mode = ... seconds

b Find the median.

Median = ... seconds

c Work out the mean.

Mean = ... seconds

4.1 Read and interpret data presented in tables

By the end of this section you will know how to:

✳ Find and understand information given in tables

Key points

✳ Use the labels on the rows and columns to find information.

✳ You may need to add values in a row or column to find a total.

✳ You may need to subtract one value from another in the table to find a difference.

1 The table shows the number of students in each year group in a secondary school.

Year group	Number of students
11	124
10	116
9	137
8	128
7	112

a How many students are in Year 11? ....

b Which is the largest year group? ....

c The senior students are in Years 10 and 11. How many senior students are there?

> **Hint**
> Look for the largest number in the second column. Find the matching year group.

> **Hint**
> Add the number of students in Year 10 to the number of students in Year 11.

....

2 Here is part of a train timetable. It shows the time that the train leaves each station.

Station	Time
Bempton	1253
Hunmanby	1303
Filey	1308
Seamer	1320
Scarborough	1325

> **Hint**
> 1253 is the same as 12:53 or 53 minutes past 12.

a What time does the train leave Hunmanby? ....

b The train is in Seamer station for 3 minutes. What time did the train arrive at Seamer?

> **Hint**
> Work out the time that is 3 minutes earlier than the time shown for Seamer.

....

25

3 The table shows the numbers of people attending a cinema on one day.
There are four screens, each showing a different film.
Each film is shown twice on the same screen.

	First show	Second show
Screen 1	462	578
Screen 2	389	547
Screen 3	536	672
Screen 4	496	758

a How many people saw the second show on Screen 2?

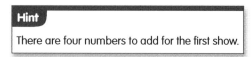

Hint
Find the number in the row for Screen 2 and the column for the second show.

....

b How many people attended the first show?

Hint
There are four numbers to add for the first show.

....

c How many people saw the most popular film?

Hint
Find the largest total for the first and second shows.

....

4 The table shows the distances in miles between some cities in England.

a Find the distance from Birmingham to London.

Hint
Follow the column down from Birmingham until it meets the row across from London.

.... miles

b Which city is 83 miles from Derby?

Hint
Look for 83 in the same row or column as Derby.
Find the other city in the same row or column as 83.

....

c Sharon drives from York to Birmingham and then from Birmingham to London.
How far does Sharon drive altogether?

Hint
Find the two distances and add them.

.... miles

Practice

5 The table shows the midday temperatures in five towns.

Location	Temperature
Cardiff	−3°C
Lancaster	−1°C
Poole	2°C
Stone	1°C
Whitby	−2°C

a What is the temperature in Whitby? ....

b Which is the coldest town? ....

c How much warmer is Poole than Whitby? ....

6 Here is part of a train timetable. It shows the time that the train leaves each station.

Station	Time
Thirsk	0610
Northallerton	0618
Darlington	0636
Thornaby	0653
Middlesbrough	0703

a What time does the train leave Thornaby? ....

b The train stops at Middlesbrough for 5 minutes. How long does it take to get from Darlington to Middlesbrough?

.... minutes

7 The table shows the numbers of cars sold by five sales people over a four-week period.

	Week 1	Week 2	Week 3	Week 4
Graham	3	4	3	4
Jinty	4	3	4	6
Matt	4	2	4	4
Sally	3	3	4	5
Mike	2	3	4	4

a How many cars did Mike sell in Week 2? ....

b How many cars were sold altogether in Week 1? ....

c Who sold the most cars? ....

d How many more cars were sold in Week 4 than in Week 3? ....

8 The table shows the distances in miles between some English cities.

	Dover	Exeter	Hull	Manchester	Sheffield	Worcester
242	Exeter					
249	284	Hull				
275	234	95	Manchester			
234	232	64	39	Sheffield		
190	133	159	112	102	Worcester	

a How far is it from Hull to Sheffield? .... miles

b Which city is less than 200 miles from Dover? ....

c Which city is closest to Hull? ....

d Kari drives from Dover to Sheffield and then from Sheffield to Hull.
   How far does she drive altogether?

.... miles

e Kari returns to Dover without going to Sheffield. How much shorter is this route?

.... miles

9 Records of rainfall in the UK go back to 1910.
   The table shows the amounts of rainfall for the wettest five years in that time.

Year	Rainfall (mm)
1954	1309
2000	1337
2002	1284
2008	1295
2012	1331

a How many of the five wettest years have occurred from 2000 onwards? ....

b Which year was the wettest? ....

c How many years are there between the two most recent entries in the table?

....

d How many years are there between the two oldest entries in the table?

....

10 Jasmine owns an ice cream parlour. The table shows how many litres of her top-selling flavours she sold each month between June and August.

	June	July	August
Belgian chocolate	126	142	135
Chocolate mint chip	94	110	102
Raspberry ripple	88	116	79
Rum and raisin	143	182	156
Vanilla	107	123	112

a Which flavour is the most popular? ....

b How much Belgian chocolate flavour ice cream was sold in July? .... litres

c In which month was the most ice cream sold? ....

d How much in total of all of these flavours was sold in June?

.... litres

e How much more Rum and raisin was sold than Vanilla over the three months?

.... litres

Need more practice ☐	Almost there ☐	Got it! ☐

Interpret charts and graphs

4.2

By the end of this section you will know how to:

✳ Read and understand information presented in charts and graphs

Key points

✳ Use the title or description of the diagram to find out what it is about.
✳ Look at the labels and try to compare values or look for a pattern.

Pictograms

1 The pictogram shows the number of cars cleaned at a car wash each day for one week.

Monday	🚗 🚗 🚗 🚗 🚗
Tuesday	🚗 🚗
Wednesday	🚗 🚗
Thursday	🚗 🚗 🚗 🚗
Friday	🚗 🚗 🚗 🚗 🚗
Saturday	🚗 🚗 🚗 🚗 🚗 🚗 🚗

Key: 🚗 = 10 cars

29

a How many cars were cleaned on Monday?

> **Hint**
>
> Each whole symbol is 10 cars and the half symbol is 5 cars.

....

b Which day was the busiest for the car wash?

> **Hint**
>
> Find the day with the longest line of symbols.

....

c What was the smallest number of cars cleaned in a day?

> **Hint**
>
> Look for the shortest line of symbols and work out the number of cars.

....

d What was the total number of cars cleaned in the week?

> **Hint**
>
> Count all of the symbols in 10s and the half symbols in 5s.

....

2 The pictogram shows how the estimated amount of cod in the North Sea has been changing since 1970.

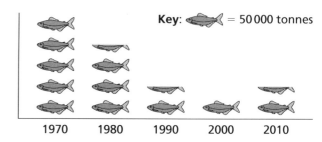

a How much cod was in the North Sea in 1970?

.... tonnes

b How much **less** cod was in the North Sea in 2000 compared with 1970? .... tonnes

c How much cod was estimated to be in the North Sea in 2010? .... tonnes

3 The pictogram shows the numbers of visitors to the top five London attractions in 2011.

British Museum	𝕏 𝕏 𝕏 𝕏 𝕏 𝕏
National Gallery	𝕏 𝕏 𝕏 𝕏 𝕏 𝕐
National History Museum	𝕏 𝕏 𝕏 𝕏 𝕏
Science Museum	𝕏 𝕏 𝕏
Tate Modern	𝕏 𝕏 𝕏 𝕏 𝕏

Key: 𝕏 = 1 million visitors

a How many people visited the Science Museum?

....

b Which was the most popular attraction?

....

c How many more people visited the National Gallery than the Tate Modern?

....

d Find the total number of visitors to the top five attractions.

....

Bar charts

4 The bar chart shows the average life spans of some popular dog breeds.

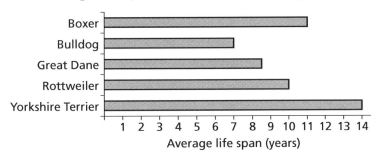

Average life span (years)

a Which breed has an average life span of 10 years?

> **Hint**
> Only one bar shows a length of 10 years.

....

b What is the average life span of a Bulldog?

> **Hint**
> Read the length of the bar for Bulldog on the life span scale.

.... years

c What is the average life span of a Great Dane? .... years

d Which breed has an average life span that is half the average life span of a Yorkshire Terrier?

> **Hint**
> Find the average life span of a Yorkshire Terrier and halve it. Match this life span to a breed.

....

5 The bar chart shows the fuel efficiencies of a sample of cars measured in miles per gallon.

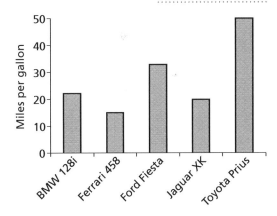

a Which car in the sample is the most fuel efficient?

....

b Which car in the sample is the least fuel efficient?

....

6 The bar chart shows the amounts of money raised by Red Nose Day for Comic Relief between 2001 and 2011.

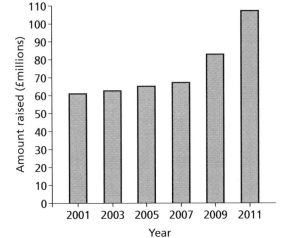

a Estimate how much money was raised in 2003 to the nearest £10 million.

£ .... million

b In which year was the amount raised more than £80 million for the first time?

....

c Which year showed the biggest increase in the amount raised?

....

31

Line graphs

Example

7 Some Year 9 students were selected at random and given a test to check their attainment levels in maths. The line graph shows the results.

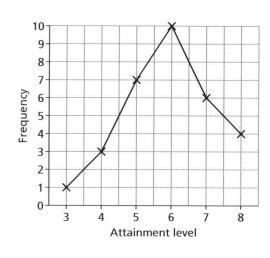

a How many students are at level 8? ....

> **Hint**
>
> Read off the height of the point for level 8.

b Which level is the mode? ....

> **Hint**
>
> The mode is the value that occurs most often.

c How many students are at level 6 or higher?

> **Hint**
>
> You need to add the frequency values for level 6 and above.

....

d How many Year 9 students were selected?

> **Hint**
>
> You need to add the frequency values for all of the levels.

....

Practice

8 A primary school teacher asked each member of her class to choose a number from 2 to 8. The line graph below on the left shows the results.

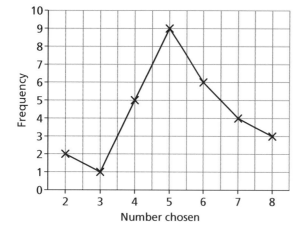
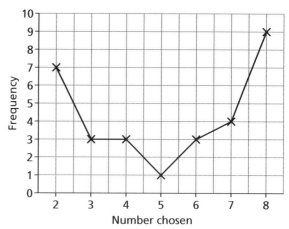

The teacher then explained that people often choose numbers towards the middle of the range and avoid numbers at the ends. She then asked the class to choose a number again. The line graph above on the right shows the second set of results.

a Which number was the mode the first time? ....

b Which number was the mode the second time? ....

c Which number was chosen with the same frequency both times? ....

d How many children are in the class?

....

e Describe the change in the results.

....

....

....

9 Charlotte recorded the number of customers in her seaside café each hour between 0900 and 1500 one day. The line graph shows this information.

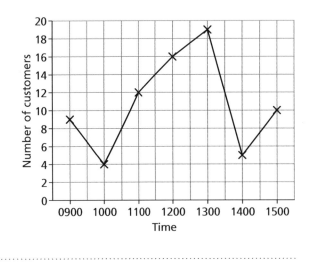

a What is the lowest number of customers shown on the graph?

....

b Explain how there may have been times between 0900 and 1500 when the number of customers was lower than this.

....

....

....

Dual bar charts

10 The dual bar chart shows how the estimated populations of red squirrels and grey squirrels in the UK have been changing since 1965.

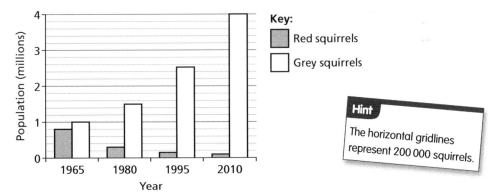

Key:
▨ Red squirrels
☐ Grey squirrels

Hint
The horizontal gridlines represent 200 000 squirrels.

a What was the estimated population of red squirrels in 1965?   *800 000*

b What was the estimated population of red squirrels in 1980?   ....

c When was the population of grey squirrels roughly five times the population of red squirrels?

Hint
Find the year in which this looks most likely and then work out the populations.

....

33

d Roughly how many grey squirrels were there for every red squirrel in 2010?

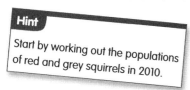

Hint
Start by working out the populations of red and grey squirrels in 2010.

....

11 The dual bar chart shows how the average prices of diesel and unleaded petrol have been changing since 2008.

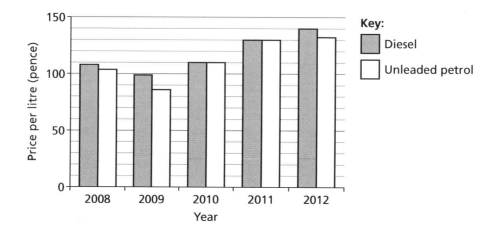

a Overall, which has been the more expensive fuel since 2008?

....

b In which years did diesel and petrol cost the same?   .... and ....

c In which year was the difference in price greatest?

....

12 The dual bar chart shows the numbers of boys and girls born at a maternity ward each week over a five-week period.

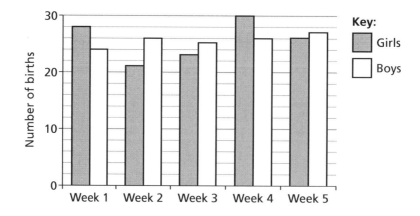

a How many girls were born in Week 3?

....

b How many more boys than girls were born in Week 2?

....

c In how many of the five weeks were more girls born than boys?

....

d In which week were the most girls born?

....

e What is the range of the total number of births in a week?

....

Pie charts

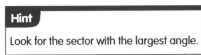

Example

13 Kelly did a survey among young adults to find out what they would most like to learn in the near future. The pie chart shows her results.

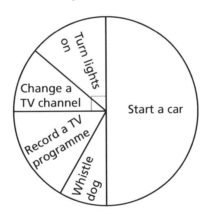

a What was the most popular response?

> **Hint**
>
> Look for the sector with the largest angle.

....

b Which response was just as popular as 'Learn to dance'?

> **Hint**
>
> Look for the sector with the same angle as 'Learn to dance'.

....

c Nine people chose 'Learn a language'.
How many people took the survey?

60° represents 9 people and 360° = 6 × 60°

so total number of people = 9 × .... = ...

Practice

14 A mobile phone company asked 84 customers to choose the most unexpected use of a mobile phone from a list of five possibilities. The pie chart shows the results.

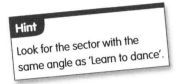

a How many customers chose 'Start a car'?

....

b How many customers chose either 'Turn lights on' or 'Change a TV channel'?

....

c Twice as many customers chose 'Record a TV programme' as 'Whistle dog'.
How many chose 'Whistle dog'?

....

Extend

15 The pie chart shows how Karen spends her wages.
Karen earns £270 per week.

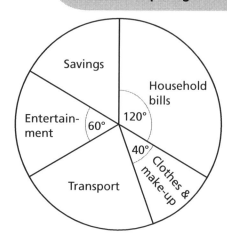

a How much does she spend on household bills?

....

b She saves as much as she spends on entertainment.
How much does she save each week?

....

c She spends twice as much on transport as she does on clothes and make-up.
How much does she spend on transport?

....

Need more practice ▢   **Almost there** ▢   **Got it!** ▢

4.3 Find totals and modes from frequency tables or diagrams

By the end of this section you will know how to:

✳ Calculate totals and find the mode of data from a frequency table or diagram

Key points

✳ You may need to add an extra row or column to a frequency table to find the total.

✳ The mode is the value that occurs with the highest frequency.

1 The table shows the total numbers of goals scored in some hockey matches.

Goals scored	1	2	3	4	5
Frequency	5	8	6	4	2

Hint
You need the total of the frequency values.

a How many matches were played?

....

b How many goals were scored altogether?

Goals scored (G)	Frequency (F)	G × F
1	5	5
2	8	16
3	6	
4	4	
5	2	
Totals		

Hint
You need to multiply each number of goals by its frequency. Add the answers.

Hint
Adding is easier in columns so it is helpful to redraw the table with an extra column.

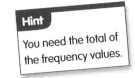

Total number of goals scored = ...

Practice

2 The table shows the numbers of people travelling per car during a survey.

Number of people (P)	Frequency (F)	
1	32	
2	21	
3	12	
4	8	
5	3	

a How many cars were observed in the survey?

....

b Find the total number of people in the cars.

....

3 The table shows the number of students in each classroom of a school during period 1 on a Monday.

Number of students	12	16	28	29	30	31	32
Frequency	3	2	7	11	14	12	9

a How many classrooms were observed?

....

b Work out the total number of students in all the classrooms.

....

Example

4 Sonia asked the students in her group to write down the names of as many Harry Potter films as they could. The table shows her results.

Number of films	1	2	3	4	5	6	7	8
Frequency	3	5	7	4	3	4	1	2

Hint

The mode is the value that occurs with the highest frequency.

What is the mode of the number of films named?

....

Practice

5 Stephen carried out a survey to find out what styles of music people like. The table shows his results.

Music style	Blues	Country	Hip hop	Jazz	Pop	Rock
Frequency	11	9	16	12	24	18

Which style is the mode?

....

6 Gavin organises a cracker eating contest.
The object is to eat as many crackers as
possible within one minute.
The results are shown in the table.

Number of crackers	Frequency
1	4
2	8
3	5
4	2
5	1

a How many people took part in the contest?

....

b How many crackers were eaten in total?

....

c What is the mode for the number of crackers eaten?

....

7 Systems of planets have now been found
orbiting stars other than our Sun.
The table shows the numbers of planets found
in orbit around stars as of January 2013.

Number of planets	Frequency
2	65
3	18
4	5
5	3
6	1
7	1
8	1

a How many stars have been found with more
than one planet in orbit?

....

b How many planets have been found in these
systems?

....

c What is the modal number of planets?

....

Don't forget!

✳ Use the .... or description of the diagram to find out what it is about.

✳ Look at the labels and try to compare values or look for a ....

✳ You may need to add an extra row or column to a frequency table to find the ....

✳ The mode is the value that occurs with the .... frequency.

Unit test

1 The table shows some information about the medals won by the top six countries in an athletics competition.

Country	Gold	Silver	Bronze	Total
United States of America	46	29	29	
People's Republic of China	38	27	23	88
Great Britain	29	17	19	
Russian Federation	24	26	32	82
Republic of Korea	13	8		28
Germany		19	14	44

a Complete the table by filling in the missing values.

b The Russian Federation had a higher medal count than Great Britain.
Explain why Great Britain is above the Russian Federation in the table.

....

....

c Find the difference in the total number of medals between the top two countries.

....

2 The table shows the distances in miles between six places in the UK.

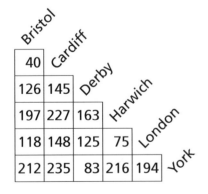

a How far is it from Cardiff to York? .... miles

b Which city is 197 miles from Harwich? ....

c Kate drives from London to Cardiff and then on to Derby.
How far does she drive altogether?

.... miles

3 The pictogram shows the numbers of visitors to a sea-life centre on five days in one week.

Monday	◯ ◖
Tuesday	
Wednesday	◯ ◯ ◯ ◖
Thursday	◯ ◯ ◯ ◯ ◯
Friday	◯ ◯ ◯ ◯ ◿
Saturday	◯ ◯ ◯ ◯ ◯ ◯ ◖

Key:
◯ represents 400 people

a How many visitors were there on Monday? ....

b There were 700 visitors on Tuesday. Complete the pictogram.

c How many more visitors were there on Thursday than on Friday?

....

4 Sports clubs in the UK reported an increase in membership since 2012. The dual bar chart shows how the membership of some clubs has changed.

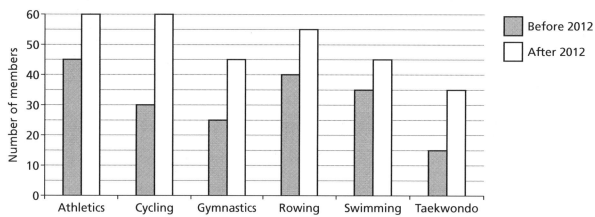

Before 2012
After 2012

a Which club had the largest increase in membership numbers? ....

b Which club more than doubled its membership? ....

c How many more people are now in the gymnastics club? ....

5 The table shows some information about the A levels studied by a group of 40 students.

Number of A levels	Frequency
1	2
2	5
3	18
4	12
5	

a Complete the table.

b What is the modal number of A levels studied? ....

c What is the range in the number of A levels studied? ....

Use and interpret a probability scale

5.1

By the end of this section you will know how to:

✳ Use the probability scale to represent probabilities

Key points

It is useful to think about an example of a probability experiment, such as rolling a dice, to understand the words used in probability.

✳ The probability experiment (in this case rolling the dice once) is called a **trial**.

✳ The result of the probability experiment (in this case, getting a score from 1 to 6) is called an **outcome**.

✳ A particular set of outcomes, such as scoring 2, 4 or 6 is an **event**.

✳ The **probability** of an event is a measure of how likely it is to happen.

✳ If an event is **impossible**, the probability that it will happen is **0**. (For example, it is impossible to get a 7 on an ordinary dice, so the probability is 0.)

✳ If an event is **certain**, the probability that it will happen is **1**. (For example, it is certain that you will get one of the numbers 1, 2, 3, 4, 5, 6 on an ordinary dice, so the probability is 1.)

✳ The probability of any event must lie on a scale from 0 to 1, including the values 0 and 1.

✳ An event that is just as likely **not** to happen as it is to happen has an **even** chance. The probability of an event with an even chance is $\frac{1}{2}$. You can also write this as 0.5 or 50%.

Example

1 Here is a probability scale.

Impossible	Unlikely	Evens	Likely	Certain
0		$\frac{1}{2}$		1

Choose the most suitable word from the probability scale to describe the probability of each of these events.

a A fair coin will land heads up when spun. .... Evens ....

> **Hint**
> The coin is just as likely to land heads up as tails up.

b It will snow in Birmingham in April. ....

> **Hint**
> The weather normally gets warmer by April, but it can get cold.

c A baby will be born today. ....

> **Hint**
> It has been estimated that a baby is born every 4 seconds.

d A cow will write a poem. ....

Practice

2 Use the same probability scale as in Question 1. Find the most suitable word to describe the probability of each of these events.

> **Hint**
> In a pack of playing cards, 26 are red and 26 are black.

a An ordinary dice is rolled and the score shown is greater than 2. ....

b A red card is selected, without looking, from a full pack of playing cards. ....

c A husband and wife like **all** of the same television programmes. ....

Example

3 The probability scale shows the positions of events A, B and C.

Write down the letter that gives the best match to the probability of each of these events. (You can use each letter more than once.)

> **Hint**
>
> Choose A if you think the event is very unlikely, choose B if you think the event has an even chance and C if you think it is very likely to happen.

a The first person to walk into a shop in London in January is wearing a coat. ....

b As you approach three different sets of traffic lights in a car, they each turn green. ....

c I pick a horse in a race because it has a funny name, and it wins. ....

d You roll an ordinary six-sided dice and score an even number. ....

Practice

4 Here is a probability scale.

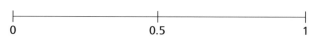

a Label the point A on the scale to show the probability that the first baby born at a maternity unit is a girl.

b Label the point B on the scale to show the probability that it will be sunny every day of the next school summer holidays where you live.

c Label the point C on the scale to show the probability that a new Olympic record will be set in some event at the next Olympic Games.

Extend

5 Here are some events.

A The next person you meet was born in a leap year.

B You will walk from Land's End to John O'Groats in one day.

C A person landing at an airport for a holiday has a suitcase to collect.

D Six girls and four boys put their names into a hat.

The first name pulled out is a girl's name.

Write A, B, C or D in each box on the probability scale to show how likely the event is to happen.

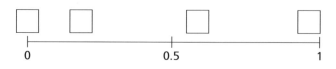

42

5.2 Write down theoretical and experimental probabilities

By the end of this section you will know how to:

* Find the theoretical probability of an event
* Find the relative frequency of an event

Key points

* If something is taken **at random** from a collection of objects, each object has the same chance of being taken, i.e. it is **equally likely**.

* When all of the possible outcomes are equally likely, the **probability** of an event is:

  Probability of an event $= \dfrac{\text{The number of outcomes in the event}}{\text{The total number of outcomes}}$

> **Hint**
>
> An outcome is something that happens in an experiment, such as a dice showing a score of 6.

* If the outcomes are not equally likely, the probability may be estimated by experiment using **relative frequency**.

* In an experiment, the **relative frequency** of an event is:

  Relative frequency of an event $= \dfrac{\text{The number of trials where the event occurred}}{\text{The total number of trials}}$

> **Hint**
>
> A trial is something that you do in an experiment, such as rolling a dice.

* The relative frequency becomes more reliable as an estimate of probability as the number of trials is increased.

Equally likely outcomes

Example

1 An ordinary six-sided dice is rolled once.

  a The probability of a score of 5 is $\dfrac{1}{\ldots}$

> **Hint**
>
> There is 1 outcome in the event. There are 6 possible outcomes.

  b The probability of an even score is ....

> **Hint**
>
> The outcomes in the event are 2, 4, 6.

  c The probability of a score less than 5 is ....

> **Hint**
>
> The outcomes in the event are 1, 2, 3, 4. Notice that 5 is not included.

  d The probability that the score is not 6 is ....

> **Hint**
>
> The outcomes in the event are 1, 2, 3, 4, 5.

  e The probability of a score of 8 is ....

> **Hint**
>
> It's impossible to score 8.

2 This five-sided spinner is spun once.

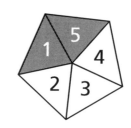

a The probability that the
spinner will land on green is ....

b The probability that the
spinner will **not** land on green is ....

c The probability that the spinner will land on an odd number is ....

d The probability that the spinner will land on an odd number on a white background is ....

e The probability that the spinner will land on an even number on a green background is ....

3 A jar contains different coloured sweets.
The table shows the number of each colour.

Colour	Red	Blue	Green
Frequency	7	8	5

A sweet is selected at random.

a The probability that the sweet is blue is $\frac{8}{.......}$

b Show your answer to part **a** on the probability scale.
Label the point A.

0   0.5   1

c The probability that the sweet is not green is ....

d Show your answer to part **c** on the probability scale.
Label the point B.

4 The table shows the results of a survey about choices of language to study at GCSE.
Each student can choose only one language.

	French	German	Spanish	Total
Boys	7	6	5	18
Girls	8	7	9	24
Total	15	13	14	42

One of these students is selected at random.

a The probability that the student is a boy is ....

b The probability that the student is a girl who chooses French is ....

c The probability that the student chooses Spanish is ....

Extend

5 The table shows some details of the cars for sale at a garage.

A car is selected at random.

	Engine size (litres)			
	1.2	**1.4**	**1.6**	**Total**
Diesel	1	2	4	7
Petrol	5	6	7	18
Total	6	8	11	25

a The probability that the car runs on diesel is ....

b The probability that the car has a 1.4 litre engine is ....

c The probability that the car has a 1.2 litre engine that runs on petrol is ....

Relative frequency

Example

6 A drawing pin can land point up or point down. Karen drops a drawing pin 50 times. The table shows her results.

	Point up	**Point down**
Frequency	31	19

a The relative frequency that the drawing pin lands point up is $\dfrac{31}{......}$

> **Hint**
> The total number of trials for this experiment is 50.

b The relative frequency that the drawing pin lands point down is ....

Practice

7 Kevin spins this spinner 40 times and records the results in a table.

Score	1	2	3	4
Frequency	8	4	5	23

a The relative frequency of scoring 1 is ....

b The relative frequency of **not** scoring 4 is ....

c The relative frequency of landing on an dark sector is ....

> **Hint**
> Combine the frequencies for the two dark sectors.

Extend

8 Megan rolls an ordinary six-sided dice 20 times. The table shows her results.

Score	1	2	3	4	5	6
Frequency	2	5	4	3	1	5

a Find the relative frequency of scoring 5. ....

b Find the relative frequency of scoring 6. ....

c Megan decides that she is much more likely to score 6 than 5. Comment on Megan's decision.

....

....

d What could Megan do to improve the reliability of her results?

....

....

List outcomes

5.3

By the end of this section you will know how to:

❋ Systematically list all of the outcomes of a single event or two successive events

Key point

❋ The **sample space** for an experiment is a list of all possible outcomes.

Example

1 The sample space for spinning a coin is ....Heads   Tails....

2 The sample space for rolling an ordinary six-sided dice is ....

> **Hint**
> List all of the outcomes for the dice.

3 Here is a spinner.

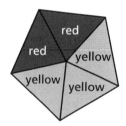

> **Hint**
> There are five outcomes for the spinner. Two outcomes are shown for you.

The sample space for the spinner is ....red   red....................................

Practice

4 Write down the sample space for this spinner.

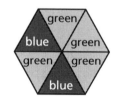

....

5 One of the colours in the Union Jack is to be selected. Write down the sample space.

....

6 Write down the sample space for this spinner.

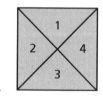

....

Example

7 A coin and an ordinary six-sided dice are used in an experiment. Write down the sample space. Use H for Head and T for Tails.

....H1   H2....................................

> **Hint**
> **H1** means Heads on the coin and **1** on the dice.

> **Hint**
> There are two outcomes for the coin and six outcomes for the dice. This means that there are 2 × 6 = 12 outcomes in the sample space.

....T1   T2....................................

8 Write down all of the possible outcomes for an experiment in which a coin is spun twice.

> **Hint**
>
> There are two outcomes for the first spin and two outcomes for the second spin, so there are $2 \times 2 = 4$ outcomes altogether.

HH

9 A car manufacturer offers three styles.

Saloon   S
Estate   E
Hatchback   H

There are three colours available.

Black   B
Red   R
White   W

> **Hint**
>
> For successive events the outcomes may be called **combinations**.

List all of the possible combinations of style and colour. The first two have been done for you.

SB   SR

10 Here are six cards.
One white card and one black card are to be selected at random.
List all of the possible combinations.
The first two have been done for you.

Q W E R X Y

QX   QY

11 Sharon has two spinners.
She spins both spinners once.
List all of the possible outcomes.
One has been done for you.

> **Hint**
>
> Write the other outcomes using brackets and a comma in the same way.

(A, 1)

Don't forget!

✳ In a probability experiment, the things that can happen are called ....

✳ The probability of an event that is **impossible** is ......

✳ The probability of any event must lie between ...... and ...... inclusive.

✳ When all of the possible outcomes are equally likely:

Probability of an event $= \dfrac{\text{The number of outcomes in} \ ....}{\text{The total} \ ....}$

✳ In an experiment:

Relative frequency of an event $= \dfrac{\text{The number of trials where the event occurred}}{..}$

✳ The relative frequency becomes more reliable as an estimate of probability as the number of trials is ....

Unit test

1 A bag contains six blue pens, three green pens and one red pen.
 A pen is selected at random.

  a Mark the point B on the scale to show the probability that the pen is blue.

  b Mark the point R on the scale to show the probability that the pen is **not** red.

  c Mark the point X on the scale to show the probability that the pen is blue, green or red.

2 The two-way table shows some information about the dogs in a rescue centre.

	Short straight hair	Long straight hair	Curly hair
Brown	5	3	1
Black	4	7	2
White	4	2	0

  a How many dogs are there at the centre? ....

  A dog is selected at random. Find the probability that the dog is:

  b brown .... **c** **not** white ....

  d black with long straight hair .... **e** white with curly hair ....

3 James has organised some maths revision sessions.
 Students choose one morning session and one
 afternoon session.
 List all of the possible combinations.
 The first one is done for you.

 SD ....

 ....

Morning	Solving equations	S
	Formulae	F
	Number	N
Afternoon	Representing data	D
	Geometry	G
	Probability	P

Metric measures

6.1

By the end of this section you will know how to:

* Use metric measures correctly
* Decide which metric unit of measure to use

Key point

* Metric units are used in many countries to measure length, weight and capacity.

Metric units of length, weight and capacity

Example

1 Connect these items to the property being measured.

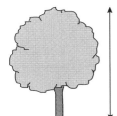
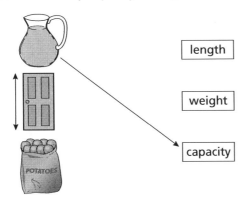

length

weight

capacity

Practice

2 Write down a suitable metric unit of measure for:

a

b

c

....

d

Paris   152

e

f

....

Extend

3 Choose a suitable estimate. Draw a ring around your choice.

    a The capacity of a coffee mug   25 ml   250 ml   2.5 litres

    b The mass of an apple   18 g   180 g   1.8 kg

    c The length of a classroom   16 mm   16 cm   16 m

Need more practice ☐   **Almost there** ☐   **Got it!** ☐

6.2 Convert between metric units

By the end of this section you will know how to:

  ✳ Convert between mm and cm, cm and metres, metres and km

  ✳ Convert between millilitres and litres, grams and kilograms

Key point

✳ Converting between metric units involves multiplying and dividing by 10, 100 or 1000.

> **Hint**
>
> 10 mm = 1 cm   100 cm = 1 metre   1000 m = 1 km
> 1000 g = 1 kg   1000 ml = 1 litre

Metres, centimetres and millimetres

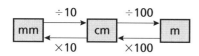

> **Hint**
>
> Centimetres are longer than millimetres, so there will be fewer of them – so divide.

Example

1 Convert these measurements by multiplying or dividing by 10 or 100.

            ÷ 10   ÷ 10   × 10
a 30 mm = ... cm   **b** 75 mm = ... cm   **c** 6 cm = ... mm

            × 10   ÷ 10   × 10
d 2.5 cm = ... mm   **e** 124 mm = ... cm   **f** 25.7 cm = ... mm

           ÷ 100   ÷ 100   ÷ 100
g 300 cm = ... m   **h** 350 cm = ... m   **i** 354 cm = ... m

           × 100   × 100   × 100
j 4 m = ... cm   **k** 4.5 m = ... cm   **l** 4.58 m = ... cm

Kilograms, kilometres and litres

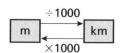 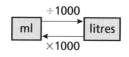 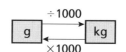

2 Multiply and divide by 1000 to convert between these units.

a 4000 m = ... km

b 4500 g = ... kg

c 4567 ml = ... litres

d 3 km = ... m

e 5.6 kg = ... g

f 2.345 litres = ... ml

g 7.2 km = ... m

h 6000 g = ... kg

i 7200 ml = ... litres

3 a Put these capacities in order. Start with the smallest amount.

250 ml   2.5 litres   0.4 litres   1800 ml   ....

b Match the equivalent lengths.

250 cm   25 cm   0.25 km   1250 m

1.25 km   2.5 m   250 mm   250 m

Need more practice	▢	Almost there	▢	Got it!	▢

Add and subtract units of measure

6.3

By the end of this section you will know how to:

✳ Add and subtract units of metric measure

Key points

✳ When adding or subtracting different units, convert one to match the other.

✳ The usual methods of addition and subtraction can be used.

1 Add or subtract these measurements.

a 45 mm + 24 mm

= .... mm

b 28 mm + 4.9 cm

= 28 mm + 49 mm

= .... mm

c 3 m 20 cm + 4 m 45 cm

= (3 + 4) m + (20 + 45) cm

= .... m .... cm

d 5 m 75 cm + 2 m 35 cm

= (............ + ...) m + (............ + ...) cm

= .... m 110 cm = ... m + 1 m 10 cm

= .... m .... cm

e 456 cm − 213 cm

= .... cm

f 6.75 m − 123 cm

= .... cm − 123 cm

= ....

51

Practice

2 Add and subtract these measurements, converting where you need to.

a 542 cm + 312 cm

..........cm

b 3000 g − 2745 g

..........g

c 32 mm + 12 mm + 56 mm

..........mm

d 4.5 km − 3.2 km

..........km

e 3.5 cm + 25 mm

..........mm

f 2 litres 500 ml + 300 ml

..........litres ....ml

g 2 m 45 cm − 75 cm

..........m ....cm

h 4 m 20 cm − 80 cm

..........m ....cm

Extend

3 a What is the total length of these two pieces of rope?

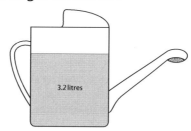

....

b What is the total distance between the library and the church?

Library $\xrightarrow{\text{750 m}}$ College $\xrightarrow{\text{1.4 km}}$ Church

....

c How much more is in the larger container?

2.6 litres

3.2 litres

....

d Add together 60 cm, 25 cm, 2 m 45 cm and 1.4 m.

....

6.4 Read scales

By the end of this section you will know how to:

✴ Read a variety of metric scales

Key point

✴ When reading scales it is important to work out how much each interval marker represents.

Example

1 How much is each interval marker worth?
Fill in the missing numbers to complete the scales.

a
```
0 2 4 6 8 10          20
```
Each interval is worth ....

b
```
0    10   20        45
```
Each interval is worth ....

c
```
0      100     200     300     400
```
Each interval is worth ....

d

Each interval is worth .... ml

e

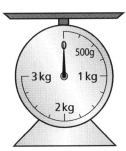

Each interval is worth .... grams

Practice

2 What measurement does each scale show?

a

.... litre .... ml

b

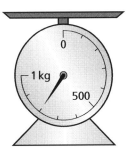

.... g

c

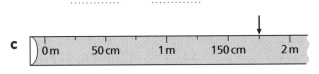

.... m .... cm

d

.... cm = ... mm

3 Draw an arrow to show the reading on each scale.

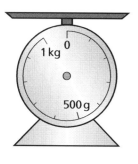

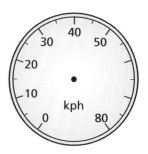

a 400 g

b 1 litre 800 ml

c 65 km per hour

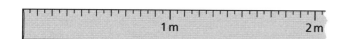

d 1 m 35 cm

e −4 °C

4 a What readings are shown? Remember to show the unit of measure.

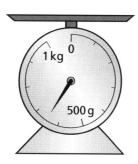

....

....

b Mark each measurement with an arrow.

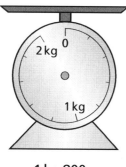

1 kg 800 g

850 ml

c Estimate the reading shown on the scale.

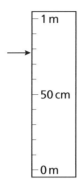

Need more practice ☐   Almost there ☐   Got it! ☐

Draw and measure lines

6.5

By the end of this section you will know how to:

✳ Draw and measure lines accurate to the nearest centimetre

Key points

✳ A metric ruler shows centimetres and millimetres.

✳ Rulers are usually 15 cm or 30 cm long.

Measuring and drawing lines

1 Measure these lines accurate to the nearest cm.

Hint

Make sure the start of the line is at the zero marker on the ruler.

a
```
cm
0    1    2    3    4
```
.... cm

b _____   **c** _____

.... cm   .... cm

2 Mark on each line the measurement shown.

A ———————————————————————— B
C ——————————————————————————— D
E —————————————————————————— F

a 5 cm from A   **b** 9 cm from C   **c** 6 cm from F

55

3 a Measure the lines *AB*, *BC* and *AC*. Which is the longest line?

A

B   *C*

AB = ... cm

BC = ... cm

AC = ... cm

Longest line is ....

4 Draw a line measuring exactly 12 cm, starting at the letter *L*. Measure to find the midpoint and mark it with the letter *M*.

L

Extend

Don't forget!

✻ Join the units with what they measure.

gram   litre   kilogram   millilitre   kilometre   cm   millimetre

capacity   distance / length   weight

✻ To convert a larger unit to a smaller unit you .... and to convert a smaller unit

to a larger unit you ....

Unit test

1 Here is a triangle *ABC*.

Measure the length of all three sides of the triangle.

AB ....   *AC* ....   *BC* ....

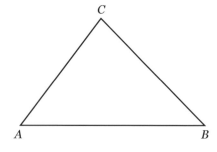

2 a Change 4000 ml into litres.   .... litres

b Change 9.5 cm into mm.   .... mm

3 Add   3 metres 45 centimetres

        2 metres 50 centimetres

           60 centimetres

.... metres .... cm

4 Write down an appropriate metric unit of measure for these items.

a The length of a table ....

b The amount of liquid in a large bucket ....

c The weight of a small bag of peanuts ....

5 a Write down the length shown.

b Write down the weight shown.

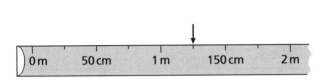

.... m .... cm

.... kg .... g

c Estimate and write down the amount shown.

.... ml

Angles and turning

7.1

By the end of this section you will know how to:
* Describe angles as turns and in degrees
* Understand clockwise and anticlockwise

Key point

* Look carefully at the starting point and the direction of the turn.

Example

1 Work out how much of a turn the minute hand of a clock moves in

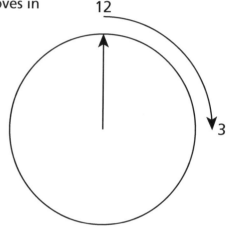

a 15 minutes

From 12 to 3 is 15 minutes, this is a $\frac{1}{4}$ turn.

b 30 minutes

From 12 to 6 is 30 minutes, this is ....

c 45 minutes

From 12 to 9 is 45 minutes, this is ....

d 1 hour

From 12 all the way round to 12 is 1 hour, this is a full turn.

Hint
A clock turns clockwise.

Other items might turn the opposite way – anticlockwise.

Athletes always run round a track in an anticlockwise direction.

Practice

2 In the diagram the arrow points North.

Work out the direction of the arrow after

Hint
For a half turn, you can turn either way.

a a half turn – the arrow points ....

b a $\frac{1}{4}$ turn anticlockwise – the arrow points ....

c a $\frac{3}{4}$ turn clockwise – the arrow points ....

d a $\frac{3}{4}$ turn anticlockwise – the arrow points ....

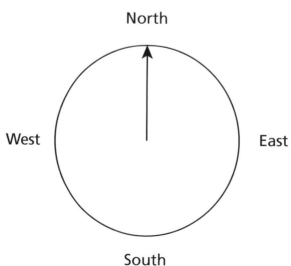

3 Write down the size and direction of each turn.

a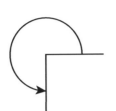

b

c

d

Extend

4 Amy faces west.

She turns to face south.

Hint

Use the diagram in Question 2 to help you.

a Write down the direction of the turn she has made. ....

b Write down the size of the turn she has made. ....

c Describe a different turn she could have made to end up facing south.

....

| Need more practice | | Almost there | | Got it! | |

Naming, measuring and drawing angles

7.2

By the end of this section you will know how to:

✳ Recognise acute, obtuse and reflex angles and right angles

✳ Use the correct notation for describing angles

✳ Draw and measure angles correct to the nearest degree

Hint

The symbol ° means degrees.

Key points

✳ An angle measures a turn.

✳ A protractor is used to measure and draw angles.

✳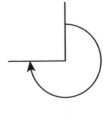

$\frac{1}{4}$ turn = 90° $\frac{1}{2}$ turn = 180° $\frac{3}{4}$ turn = 270°   full turn = 360°

✳ This table shows different types of angles.

Acute angle	Right angle	Obtuse angle
Less than 90°	Exactly 90°	Between 90° and 180°

Straight line		Reflex angle
Exactly 180°		Between 180° and 360°

1 Write acute, obtuse or reflex next to these angles.

Hint
More than $\frac{3}{4}$ of a turn – almost 360°.

a   b   c   d

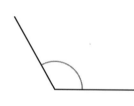 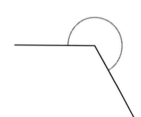 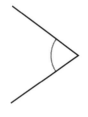 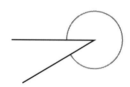

2 Write acute, obtuse or reflex next to these angles.

a   b   c   d

 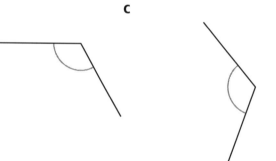 

3 What type of angle is this?

Describe it using the capital letter notation.

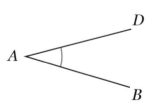

Angle DAB is an acute angle.

Hint
The point of the angle is the middle letter.
You could also have named it angle *BAD*.

Practice

4 Describe each angle using letters and say what type of angle it is in each case.

a

b

c

d

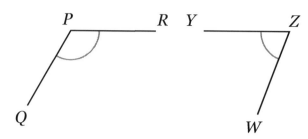

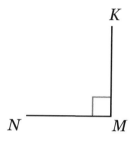

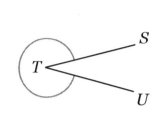

Extend

5 The diagram shows an arrowhead.

It has 7 interior angles (shown).

a How many acute angles are there?

Name them using letters.

....

....

....

....

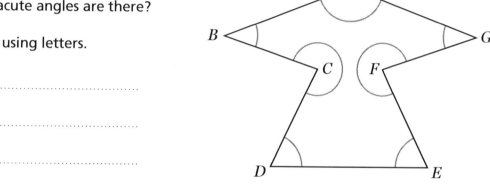

b How many obtuse angles are there?

Name them using letters.

....

....

c How many reflex angles are there?

Name them using letters.

....

....

Example

6 Measure each angle carefully.

a

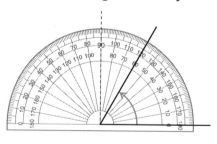

.... °

b

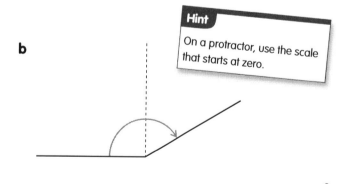

> **Hint**
>
> On a protractor, use the scale that starts at zero.

.... °

61

Practice

7 Measure these angles carefully.

a

b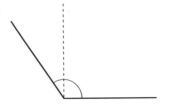

c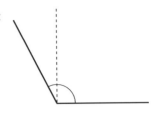

.............................° ....° ....°

d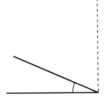

e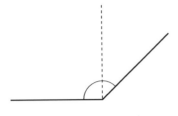

f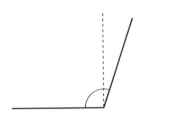

.............................° ....° ....°

Extend

8 Measure the angles in this triangle.

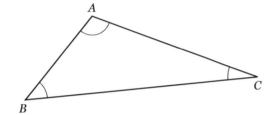

$A =$ ....° $B =$ ....° $C =$ ....°

Hint

Position the centre of the protractor on the dot, with the zero line along the line drawn. Use the scale that starts at zero.

Drawing angles

Example

9 Draw and label these angles using a protractor. Start at the point shown.

a 80°

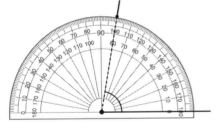

b 32°

Practice

10 Draw these angles using a protractor. Start at the point shown.

a 42°

b 143°

c 162°   **d** 48°

Extend

11 a Measure the line *PQ* and mark the mid-point *M* on it.

P ——————————————————— *Q*

b At *M* draw and label an angle of 75°.

c Measure and label the other angle at *M*.

Don't forget!

✳ An angle is a measure of turn and can be measured in degrees.

✳ A protractor is used to measure and draw angles.

✳ Angles can be described using capital letters, e.g. angle *BAC*.

✳ An .... angle is less than 90°; an .... angle is between 90° and 180°
and a .... angle is between 180° and 360°.

Unit test

1 Write acute, obtuse or reflex next to these angles.

a   **b**   **c**   **d**

      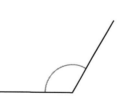   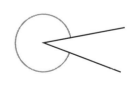

....

2 What types of angles are these? (acute, obtuse etc.)

Describe them using the capital letter notation.

a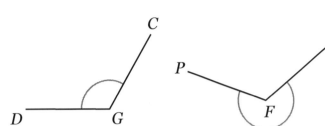

b

c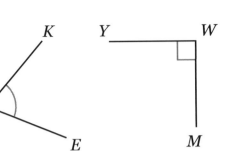

d

....

3 Measure these angles using a protractor.

a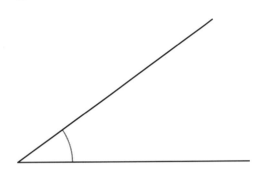

b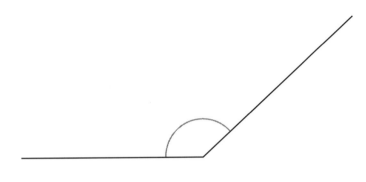

....

4 Draw these angles using a protractor.

a 105°

_____

b 72°

_____

8.1 Angles on a straight line and angles round a point

By the end of this section you will know how to:

✳ Calculate angles on a straight line

✳ Calculate angles round a point

✳ Recognise vertically opposite angles

Key points

✳ A straight line, or half turn = 180°, so angles on a straight line add up to 180°.

✳ A full turn = 360°, so angles round a point add up to 360°.

Example

1 Work out the size of angle a.

$a + 50° = 180°$

$a = 180° -$ ....°

angle $a =$ ....°

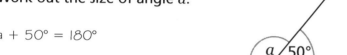

> **Hint**
> Lower case letters are often used to indicate angles.

2 Work out the size of angle b.

............° $+ 60° + b =$ ....°

............° $+ b =$ ....°

$b =$ ....° $-$ ....°

angle $b =$ ....°

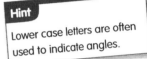

> **Hint**
> Angles on a straight line add up to 180°.

Practice

3 Work out the size of angle c.

angle $c =$ ....°

4 Work out the size of angle d.

angle $d =$ ....°

5 Work out the size of angle e.

angle $e =$ ....°

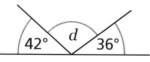

> **Hint**
> This diagram has a right angle in it.

6 Work out the size of angle f.

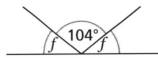

The angles marked f are equal.

angle $f =$ ....°

7 Work out the size of angle g.

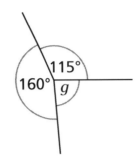

$g + 160° + 115° = 360°$

$g + 275° = 360°$

$g = 360° -$ ....°

angle $g =$ ....°

Hint

Angles round a point add up to 360°

Hint

When two straight lines intersect like this, angles such as h are equal. These angles are known as **vertically opposite** angles. What does that tell you about angle j and 134°?

8 Work out the sizes of angles h and j.

angle $h =$ ....°

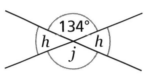

angle $j =$ ....°

9 Work out the size of angle k.

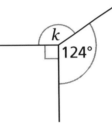

angle $k =$ ....°

10 Work out the size of angle m.

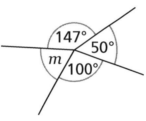

angle $m =$ ....°

11 Work out the size of angle n.

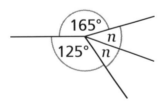

angle $n =$ ....°

Triangle properties

8.2

By the end of this section you will know how to:

✳ Identify different types of triangles

✳ Work out the sizes of angles in triangles

Key points

✳ In a triangle, the three angles add up to 180°.

✳ This table shows different types of triangle.

Right-angled	Scalene	Isosceles	Equilateral
One angle is exactly 90°.	All three sides are different lengths. All three angles are different sizes.	Two sides are equal in length. Two angles are the same size.	All three sides are equal in length. All three angles are 60°.
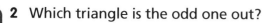			

Example

1 Which triangle is the odd one out?

Give a reason for your answer.

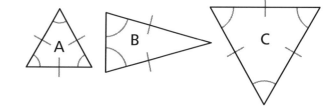

Triangle B, because A and C are both equilateral, B is isosceles.

Practice

2 Which triangle is the odd one out?

Give a reason for your answer.

3 Draw sketches of these triangles.

a An isosceles triangle   **b** A scalene triangle   **c** A right-angled triangle

Hint

Use a ruler.

Extend

4 In this diagram, identify the triangles described by giving the three letter name (e.g. triangle *PQR*).

a An equilateral triangle.

Triangle ....

b A right-angled triangle.

Triangle ....

c A scalene triangle.

Triangle ....

d An isosceles triangle.

Triangle ....

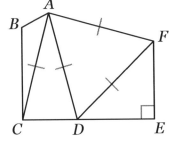

Example

5 Work out the size of angle *m*.

m + 46° + 63° = 180°

m + 109° = 180°

m = 180° − ....°

angle *m* = ...°

Hint

The three angles in a triangle add up to 180°

Practice

6 Work out the size of angle *n*.

angle *n* = ...°

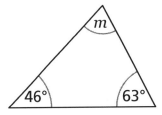

7 Work out the size of angle *p*.

angle *p* = ...°

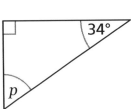

Extend

8 Work out the size of angle *t*.

angle *t* = ...°

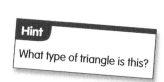

Hint

What type of triangle is this?

Don't forget!

✳ A straight line, or half turn = ...°, so angles on a straight line add up to ....°

✳ A full turn = ...°, so angles round a point add up to ....°

✳ In a triangle, the three angles add up to 180°.

Unit test

1 Work out the size of angle g.

angle g = ... °

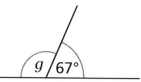

2 Work out the size of angle h.

angle h = ... °

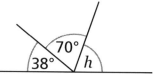

3 Work out the size of angle p.

angle p = ... °

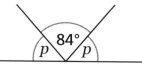

4 Work out the size of angle w.

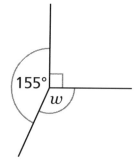

angle w = ... °

5 Work out the sizes of angles x and y.

angle x = ... °

angle y = ... °

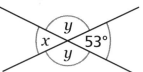

6 Work out the size of angle z.

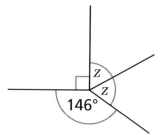

angle z = ... °

7 Which triangle is the odd one out?

Give a reason for your answer.

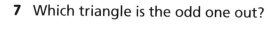

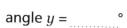

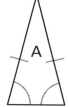

8 Work out the size of angle k.

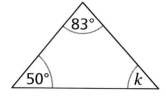

angle k = ... °

9 Work out the size of angle y.

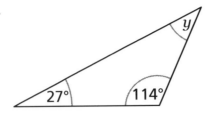

angle y = ...°

10 Work out the size of angle a.

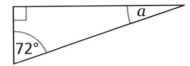

angle a = ...°

11 Work out the size of angle m.

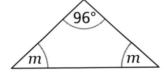

angle m = ...°

12 Work out the size of angle c.

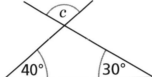

angle c = ...°

Accurate drawings

9.1

By the end of this section you will know how to:

* Draw triangles accurately when at least one angle is given
* Draw triangles accurately when given all three sides

Key points

* Use a metric ruler to measure lines accurately.
* Use a protractor to measure angles accurately.
* Use a pair of compasses to locate the point of intersection of two sides of a triangle.

Example

1 **a** Make an accurate drawing of this triangle.

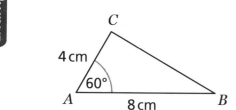

Hint
This is just a sketch, you need to draw it full size.

Hint

Step 1: Draw a line *AB*, 8 cm long.

Step 2: Measure an angle of 60° at *A*.
Position the centre of the protractor at *A*, with the zero line along *AB*.
Use the scale that starts at zero.

Step 3: Measure 4 cm from *A* along the 60° line. Label this point *C*.
Join *BC*.

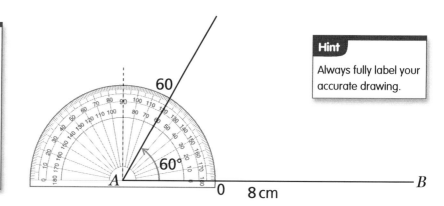

Hint
Always fully label your accurate drawing.

b Measure the length of *BC*   *BC* = ... cm

Practice

2 **a** Make an accurate drawing of this triangle.

Hint
The line *DE* has been drawn for you.

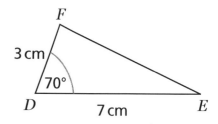

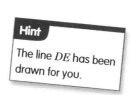

b Measure the length of *EF*. $EF = $ .... cm

3 a Make an accurate drawing of this triangle.

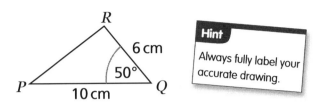

> **Hint**
> Always fully label your accurate drawing.

b Measure the length of *PR*.

$PR = $ .... cm

4 a Make an accurate drawing of this triangle.

> **Hint**
> Step 2: Measure an angle of 40° at *J*.
> Position the centre of the protractor at *J*, with the zero line along *JK*.
> Use the scale that starts at zero.

> **Hint**
> Step 1: Draw a line *JK*, 7.5 cm long.

> **Hint**
> Step 3: Measure an angle of 60° at *K*.
> Step 4: Draw the lines *JL* and *KL* long enough so that they intersect.

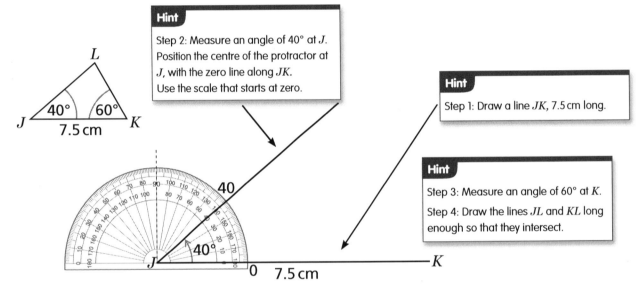

b Measure the lengths of *JL* and *KL*.

$JL = $ .... cm $KL = $ .... cm

5 a Make an accurate drawing of this triangle.

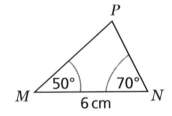

b Measure the lengths of *MP* and *NP*.

MP = ... cm

NP = ... cm

6 a Make an accurate drawing of this diagram.

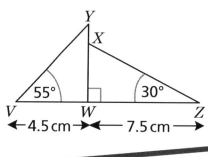

Hint
Start by drawing the line *VWZ* and measure a right angle at *W*.

b Measure the lengths of *VY*, *XZ* and *XY*.

VY = ... cm

XZ = ... cm

XY = ... cm

7 a Make an accurate drawing of this triangle.

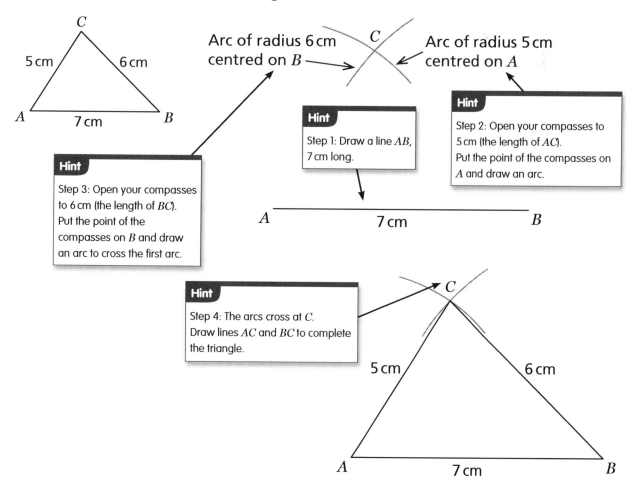

Arc of radius 6 cm centred on *B* →

Arc of radius 5 cm centred on *A*

Hint
Step 1: Draw a line *AB*, 7 cm long.

Hint
Step 2: Open your compasses to 5 cm (the length of *AC*).
Put the point of the compasses on *A* and draw an arc.

Hint
Step 3: Open your compasses to 6 cm (the length of *BC*).
Put the point of the compasses on *B* and draw an arc to cross the first arc.

Hint
Step 4: The arcs cross at *C*.
Draw lines *AC* and *BC* to complete the triangle.

b Measure the size of angle *BAC to the nearest degree.*

angle *BAC* = ... °

8 a Make an accurate drawing of this triangle.

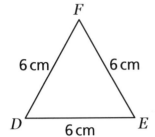

D _____ 6 cm _____ *E*

b What do you notice about the sizes of all three angles? ....

c What type of triangle is this? ....

9 a Make an accurate drawing of this triangle.

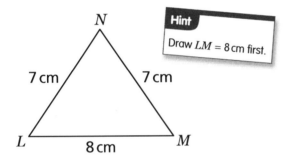

b Measure the sizes of angles *NLM* and *LMN to the nearest degree.*

angle *NLM* = ... °

angle *LMN* = ... °

c What type of triangle is this? ....

Extend

10 a Make an accurate drawing of this triangle.

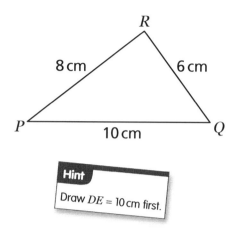

Hint

Draw $DE = 10$ cm first.

b Measure the sizes of angles RPQ, PQR and PRQ.

angle $RPQ =$ ....°

angle $PQR =$ ....°

angle $PRQ =$ ....°

c What type of triangle is this? ....

Unit test

1 **a** Make an accurate drawing of this triangle.

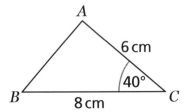

b Measure the length of *AB*.

AB = ... cm

2 **a** Make an accurate drawing of this triangle.

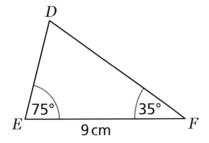

b Measure the lengths of *DE* and *DF*.

DE = ... cm

DF = ... cm

c What type of triangle is this? ....

3 a Make an accurate drawing of this triangle.

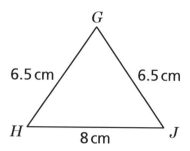

b Measure the sizes of angles *GHJ* and *GJH*.

angle *GHJ* = ... °

angle *GJH* = ... °

c What type of triangle is this? ....

4 a Make an accurate drawing of this triangle.

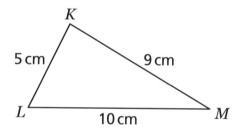

b Measure the sizes of angles *LKM*, *KLM* and *LMK*.

angle *LKM* = ... °

angle *KLM* = ... °

angle *LMK* = ... °

Reflection

10.1

By the end of this section you will know how to:

✳ Carry out simple reflections of shapes

Key point

✳ When an object is reflected, the mirror line is a line of symmetry between object and image.

Example

1 Reflect these shapes in the mirror line (shown dashed).

a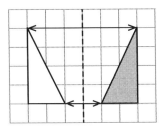

Hint
The shaded triangle image is exactly the same size and the same distance from the mirror line.

b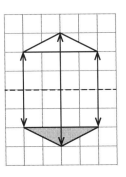

Hint
When an object is reflected, the image is on the other side of the mirror and faces in the opposite direction.

c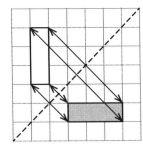

Hint
You can work out the image positions one point at a time and then draw the final image.
Draw 'diagonals' to find the image points.

Practice

2 Reflect these shapes in the mirror line.

a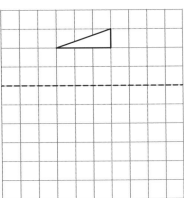

b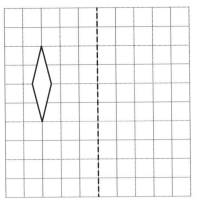

c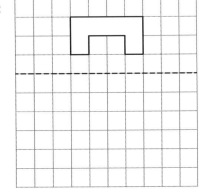

3 Reflect these shapes in the mirror line.

a

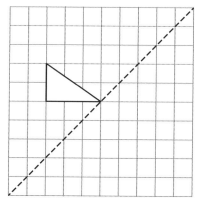

b

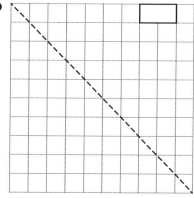

Hint

Reflect one point at a time for each of these.
Draw 'diagonals' to find the image points.

4 Reflect these shapes in the mirror line.

Extend

a

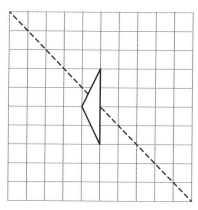

b

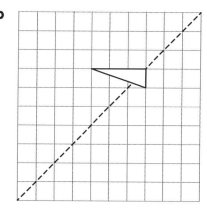

Need more practice ☐   **Almost there** ☐   **Got it!** ☐

Enlargement

10.2

By the end of this section you will know how to:

✳ Identify simple enlargements of shapes

Key point

✳ When an object is enlarged, the object and image are the same shape but different sizes.

Hint

To find the scale factor, you compare the lengths of corresponding sides. (You could compare the heights of the triangles instead.)

Example

1 The grids show a shape and its enlargement (shaded).
What is the scale factor of the enlargement in each case?

Hint

An enlargement changes the size of the object but keeps it the same shape.

a

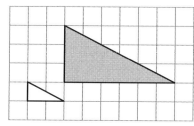

The base of the small triangle is 2 squares long.
The base of the shaded, large triangle is 6 squares long.
6 ÷ 2 = 3, so scale factor = 3

b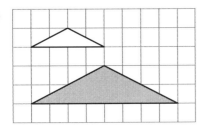

The base of the small triangle is 4 squares long.
The base of the shaded, large triangle is

.... squares long.

.... ÷ 4 = ... ,

so scale factor = ...

2 The grids show a shape and its enlargement (shaded).
What is the scale factor of the enlargement in each case?

a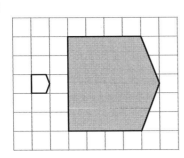

b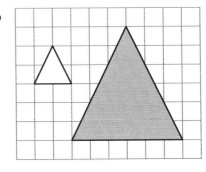

Scale factor = ...   Scale factor = ...

3 Shape A is an enlargement of shape a. Write down the scale factor of enlargement for each pair of shapes.

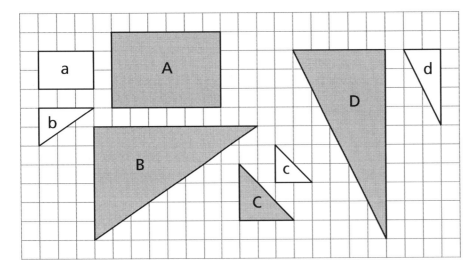

Hint

Work out the multiplier in each case.

Scale factor   **a** ....   **b** ....   **c** ....   **d** ....

Congruent shapes and similar shapes

10.3

By the end of this section you will know how to:

* Recognise shapes that are congruent
* Recognise shapes that are similar

Key points

* Congruent shapes are shapes that are identical.
* Similar shapes are different in size, but have equal angles.
* For similar shapes, the scale factor is the same for all pairs of corresponding sides.

Example

1 This shape has been reflected in the mirror line.

Are these shapes congruent?

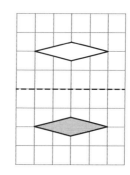

> **Hint**
> Are the shapes identical? If 'yes', then they are congruent.

Yes. The shapes are identical, so they are congruent.

Practice

2 Which of these pairs of shapes are congruent?

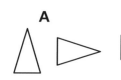 A   B   C   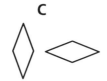 D   E   F

Example

3 The grid shows a shape and its enlargement by scale factor 2.
Are the shapes congruent or similar?

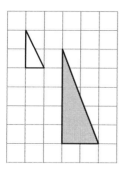

> **Hint**
> Are the shapes identical? If 'yes', then they are congruent.
> Are all the pairs of corresponding sides linked by the same scale factor? Do they have equal angles? If 'yes', then they are similar.

The shapes are not identical, so they .... congruent.

Each length in the image is double the corresponding length in the object and the angles are

the same in both shapes, so the shapes .... similar.

Practice

4 Are these shapes similar?

Hint

Measure the lengths of pairs of sides.

Hint

AB and PQ are corresponding sides.
BC and QR are corresponding sides.
AC and PR are corresponding sides.

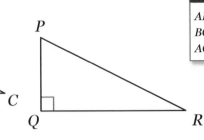

Hint

What do you notice about these scale factors?

$AB = 1\,cm$ $PQ = 2\,cm$ so $PQ =$ .... $\times AB$

$BC =$ .... cm $QR =$ .... cm   so $QR =$ .... $\times BC$

The shapes .... similar.

5 a Which of these pairs of shapes are similar?

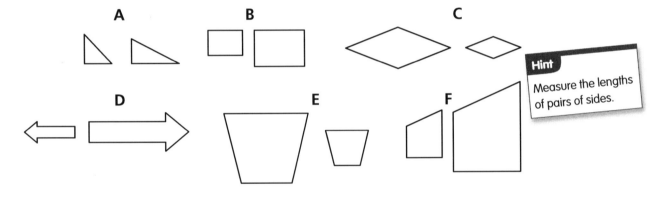

A   B   C

D   E   F

Hint

Measure the lengths of pairs of sides.

....

b Write down the scale factor of the enlargement for each similar pair.

....

Don't forget!

✳ When an object is reflected, the mirror line is .... between object and image.

✳ When an object is enlarged, the object and image are the same shape but different ....

✳ Congruent shapes are shapes that are ....

✳ Similar shapes are different in size, but have equal ....

✳ For similar shapes, the scale factor is the .... for all pairs of corresponding sides.

Extend

Unit test

1 Reflect these shapes in the mirror line.

a

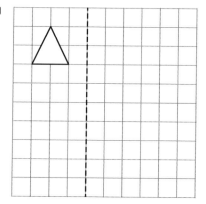

b

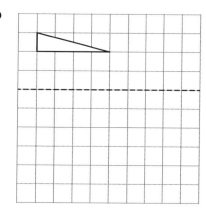

c

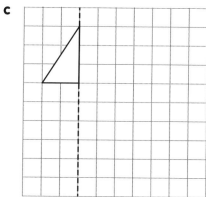

d

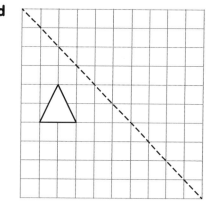

e

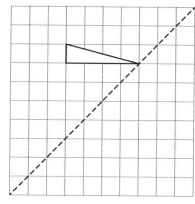

f

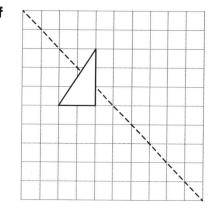

2 The grids show a shape and its enlargement (shaded).
What is the scale factor of the enlargement in each case?

a

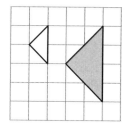

b

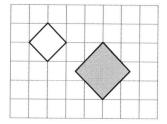

c

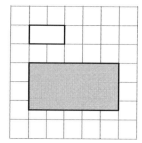

Scale factor = ...

Scale factor = ...

Scale factor = ...

3 Shape A is an enlargement of shape a. Write down the scale factor of enlargement for each pair of shapes.

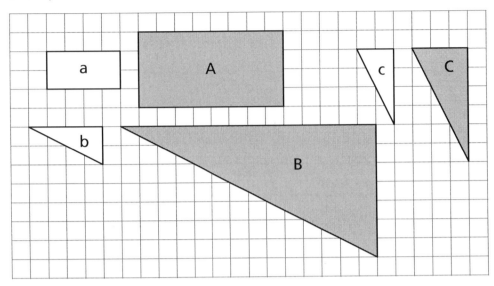

Scale factor **a** .... **b** .... **c** ....

4 Which of these pairs of shapes are congruent?

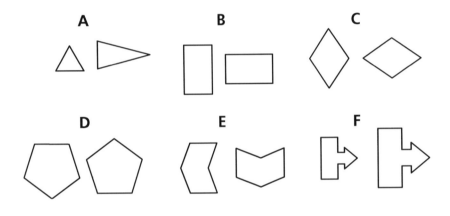

A   B   C

D   E   F

....

5 a Which of these pairs of shapes are similar?

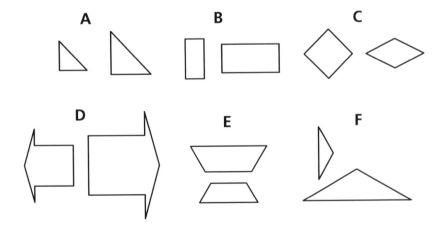

A   B   C

D   E   F

....

b If a pair is similar, write down the scale factor of the enlargement.

....

11.1 Circles

By the end of this section you will know how to:

* ✳ Draw a circle of a given radius or diameter
* ✳ Identify some of the component parts of a circle

Key points

* ✳ **Diameter**: A straight line passing through the centre.
* ✳ **Radius**: A straight line from the centre to the edge of the circle.
* ✳ **Circumference**: The distance around the edge of the circle.
* ✳ **Tangent**: A straight line that just touches the circle at one point only.
* ✳ **Chord**: A line segment joining two points on the edge of the circle.
* ✳ **Arc**: A part of the edge of the circle.

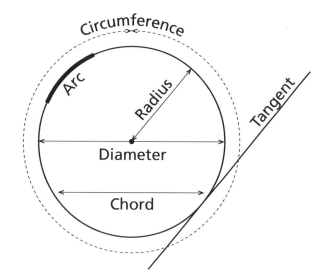

Example

1   **a**   Use a pair of compasses to draw a circle of diameter 4 cm.

     b   Mark the centre of the circle.

     c   Draw and label a radius.

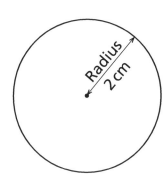

       i   Diameter = 4 cm
          Radius = Diameter ÷ 2 = 4 ÷ 2 = 2 cm
          Open your compasses to a radius of 2 cm and draw the circle.

       ii   Mark the centre.

       iii   Join the centre to any point on the edge. This is a radius.

Practice

2 **a** Use a pair of compasses to draw each of these circles in the space below.

    i Radius 2 cm

    ii Diameter 6 cm

  b On each circle, draw

    i a diameter

    ii a chord

    iii a tangent

3 **a** Use a pair of compasses to draw a circle of diameter 5 cm.

  b Mark two points on the circumference.
Join them with a line along the circumference. What is this line called?

....

4 a Use a pair of compasses to draw a circle of radius 3 cm.

   b Mark two points, A and B, on the circumference.
         What is the name of the part of the circumference joining A and B?

   ....

Extend

5 a Use a pair of compasses to draw a circle of radius 3.5 cm. Mark the centre.

   b Choose a point, P, on the circumference.
         Draw a straight line from the centre to P.

   c Draw a tangent at point P.

Qualateral... Quadrilaterals

By the end of this section you will know how to:

✳ Identify quadrilaterals from their properties

Key points

✳ Parallel lines never meet. They are the same distance apart and slope in the same direction. They are indicated by arrows on the lines.

✳ Perpendicular lines cross at right angles (90°). A right angle is indicated by a small square angle symbol.

✳ Bisect means to divide into two equal parts.

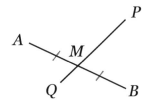

Line *PQ* bisects line *AB*.

$AM = MB$

M is the mid-point of *AB*.

✳ A quadrilateral has 4 sides.

Square		• 4 equal sides • 4 right angles
Rectangle		• opposite sides equal • 4 right angles
Parallelogram		• opposite sides equal and parallel • opposite angles equal • diagonals bisect each other
Trapezium		• one pair of parallel sides
Rhombus		• 4 equal sides • opposite sides parallel • opposite angles equal • diagonals bisect each other at right angles
Kite		• 2 shorter sides equal • 2 longer sides equal • one pair of opposite angles equal • diagonals cross at right angles

Example

1 Which quadrilateral has 4 equal sides but no right angles?

Two quadrilaterals have 4 equal sides – a square and a rhombus.
A square has 4 right angles (90°), so the quadrilateral required cannot
be a square. So, the quadrilateral described is a rhombus.

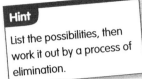

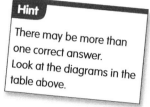

Hint
List the possibilities, then work it out by a process of elimination.

Practice

2 Work out which quadrilateral is being described.

Hint
There may be more than one correct answer.
Look at the diagrams in the table above.

a Two pairs of equal sides and 4 right angles.

....

b One pair of parallel sides.

....

c Two pairs of opposite angles equal and opposite sides parallel.

....

d Two pairs of opposite angles equal and opposite sides parallel, but no right angles.

....

e One pair of opposite angles equal.

....

Extend

3 Here is a right-angled triangle.

Draw diagrams to show how two of these right-angled triangles can be joined together to make

a a rectangle

b a kite

c an isosceles triangle.

Don't forget!

* A diameter is a straight line passing through the .... of the circle.

* A .... is a straight line from the centre to the edge of the circle.

* The circumference is the distance around the .... of the circle.

* A tangent is a straight line that just .... the circle.

* A .... is a line segment joining two points on the edge of the circle.

* An arc is a part of the .... of the circle.

* Quadrilaterals have .... sides.

* Learn the main properties of each quadrilateral.

Unit test

1   Here is a diagram of a circle, centre O.

   Name these features.

   AC is a ....

   OB is a ....

   AD is a ....

   PQ is a ....

   From C to D is an ....

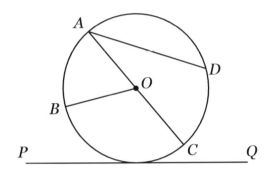

2 **a** Use a pair of compasses to draw a circle of radius 2.5 cm. Mark the centre.

   b Choose a point, P, on the circumference.
       From point P, draw a diameter. Label the other end of the diameter Q.

   c Draw a tangent at point Q.

3 Which quadrilateral is being described? (There may be more than one correct answer.)
   The two diagonals are equal in length and they bisect each other at right angles.

   ....

4 Here is a scalene triangle.
   Draw a diagram to show how two of these scalene triangles can be joined
   together to make a parallelogram.

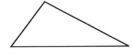

Perimeter of rectangles

12.1

By the end of this section you will know how to:

✳ Work out the perimeter of rectangles

Key points

✳ The perimeter of a shape is the distance all the way round its edge.

✳ To find the perimeter of a rectangle, add together the lengths of all the edges.

Example

1 Write down the lengths of the sides of these rectangles. Then add them together to find the perimeter.

a
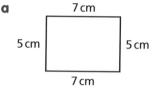
7 cm, 5 cm, 5 cm, 7 cm

b
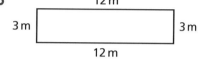
12 m, 3 m, 3 m, 12 m

c

14 cm, 26 cm

a 5 cm + 7 cm + ... cm + ... cm = ... cm

b 12 m + ... m + ... m + ... m = ... m

c .... cm + ... cm + ... cm + ... cm = ... cm

Practice

2 Work out the perimeter of these rectangles.

> **Hint**
> Opposite sides of a rectangle are equal.

a
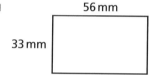
56 mm, 33 mm

b
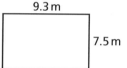
9.3 m, 7.5 m

Perimeter = ... mm   Perimeter = ... m

3 Work out the perimeter of these squares.

> **Hint**
> A square is a special type of rectangle.

a
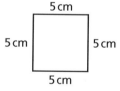
5 cm, 5 cm, 5 cm, 5 cm

b

7.4 cm

Perimeter = ... cm   Perimeter = ... cm

4 Use the formula to find the perimeter of these shapes.

> **Hint**
> The formula for the perimeter of a rectangle is:
> 2 × length + 2 × width

a
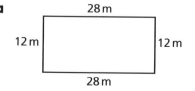
28 m, 12 m, 12 m, 28 m

b
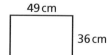
49 cm, 36 cm

Perimeter = ... m   Perimeter = ... cm

Extend

5 Work out the length of the missing side.

a

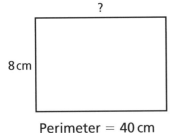

Perimeter = 40 cm

.... cm

b A square has a perimeter of 24 cm. What is the length of each of its sides?

.... cm

Perimeter of L shapes

Example

6 Here is a shape made from rectangles. Work out the lengths of the missing sides.

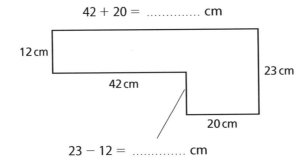

42 + 20 = ... cm

12 cm

23 cm

42 cm

20 cm

23 − 12 = ... cm

Now work out the distance all the way round the shape. This is the perimeter:

.... + ... + ... + ... + ... + ... = ... cm

Practice

7 Find the perimeter of this shape.

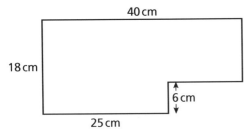

40 cm

18 cm

6 cm

25 cm

Hint

First find the lengths of any missing sides.

Perimeter = ... cm

Extend

8 This shape is made up of four identical squares.
Find the perimeter of the shape.

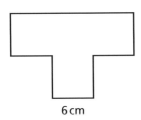

6 cm

Perimeter = ... cm

12.2 Area of rectangles

By the end of this section you will know how to:

✳ Work out the area of rectangles

Key points

✳ The area of a shape is the amount of space inside it.

✳ To find the area of a rectangle multiply the width by the length.

✳ To find the area of a shape made from rectangles, add the areas of the individual rectangles.

Example

1 Find the areas of these rectangles by counting the squares.
Multiply the two lengths together to check your answer.

> **Hint**
> Area is recorded in square units, e.g. mm², cm², m².

a

3 cm

2 cm

b

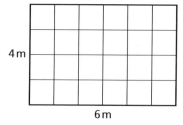

4 m

6 m

c

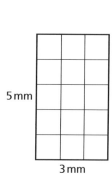

5 mm

3 mm

area = ... cm²

3 × 2 = ... cm²

area = ... m²

.... × .... = ... m²

area = ... mm²

.... × .... = ... mm²

Practice

2 Find the areas of these rectangles using the formula.

a

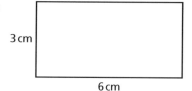

3 cm

6 cm

b

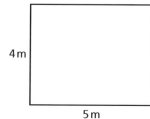

4 m

5 m

c

2 mm

3 mm

> **Hint**
> Area of rectangle = width × length.

Area = ... cm²

Area = ... m²

Area = ... mm²

Extend

3 a Work out the total area of this shape.

3.6 m

4.8 m

4 m

2.5 m

Area = ... m²

b Work out the shaded area of this shape.

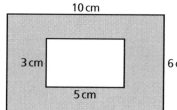

10 cm

3 cm

5 cm

6 cm

> **Hint**
> Find the difference between the areas of the larger and the smaller rectangles.

Area = ... cm²

Volume of cuboids

12.3

By the end of this section you will know how to:

∗ Find the volume of a cuboid using the formula:
  volume = length × width × height

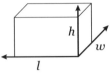

h

w

l

3 dimensions

Key points

∗ **Volume** is the space a 3D shape takes up.

∗ **3D** means 3 dimensions: length (l), width (w), height (h).

∗ A cuboid is a 3D shape with three pairs of equal rectangular faces.

Example

1 Count the cubes to find the volume of each shape.

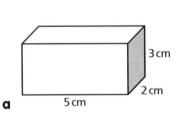

a

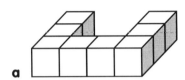

b

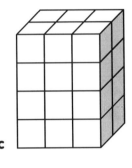

c

Hint

Volume is recorded in cubic units, e.g. mm³, cm³, m³.

$3 \times 2 \times 2 = $ ....

.... × .... × .... = ...

volume = ... cm³

volume = ... cm³

volume = ... cm³

Practice

2 Find the volume of each cuboid by using the formula.

Hint

Volume of cuboid = length × width × height.

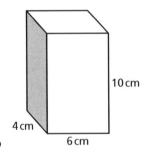

a 5 cm   2 cm   3 cm

b 6 cm   4 cm   10 cm

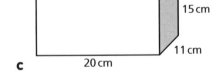

c 20 cm   11 cm   15 cm

Volume = ...

Volume = ...

Volume = ...

= .... cm³

= .... cm³

= .... cm³

3 Find the volume of each cuboid by using the formula.

a 45 mm   55 mm   35 mm

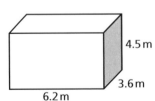

b 6.2 m   3.6 m   4.5 m

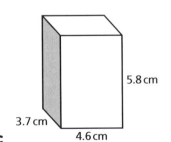

c 3.7 cm   4.6 cm   5.8 cm

Volume = ...

Volume = ...

Volume = ...

= .... mm³

= .... m³

= .... cm³

Cubes

Extend

4 a Find the volume of each cube. Record your answers in the table.

Hint

A cube is a special type of cuboid. All dimensions are equal in a cube.

Length of edges	l	w	h	Volume
1 cm	1	1	1	1 cm³
2 cm	2	2	2	
3 cm	3			
4 cm	4			
5 cm	5			
6 cm	6			
7 cm	7			
8 cm	8			
9 cm	9			
10 cm	10			

b A cube has a volume of 125 cm³.
What is the length of each edge?

.... cm

c A cube has a volume of 1000 m³.
What is the length of each edge?

.... m

Don't forget!

✳ To find the perimeter of a rectangle ....

✳ To find the area of a rectangle ....

✳ To find the volume of a cuboid ....

✳ A cube is a special type of cuboid because ....

Unit test

1 Here is a shaded rectangle drawn on a grid of centimetre squares.

a Find the perimeter of the rectangle.

.... cm

b Find the area of the rectangle.

.... cm²

2 Here is a shaded shape drawn on a grid of centimetre squares.

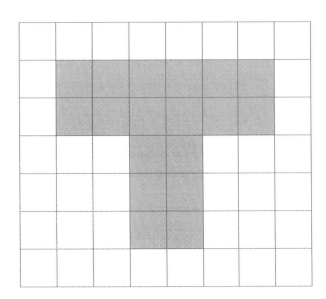

a Find the perimeter of the shape.

.... cm

b Find the area of the shape.

.... cm²

3 Here is a shape made from rectangles.

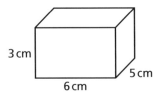

12.6 cm

3.5 cm

5.9 cm

8.4 cm

Work out the perimeter of the shape.

.... cm

4 Work out the volume of this cuboid.

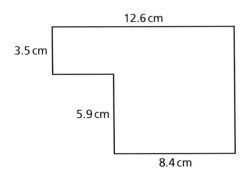

3 cm

6 cm

5 cm

.... cm³

5 Work out the volume of this cube.

4 m

.... m³

6 Work out the volume of this cuboid.

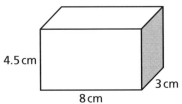

4.5 cm

8 cm

3 cm

.... cm³

Statistics test

Time: 60 minutes

Edexcel publishes Sample Assessment Material on its website. This Statistics test has been written to help you practise what you have learned and may not be representative of a real exam paper.

1 Helen owns a hair salon.
The pictogram shows the numbers of clients the salon had on five days in one week.

a There were 36 clients on Wednesday.
Show this information on the pictogram. (1)

b How many clients were there altogether in the week?

Monday	𝔁𝔁𝔁𝔁𝔁
Tuesday	𝔁𝔁𝔁𝔁
Wednesday	
Thursday	𝔁𝔁𝔁𝔁𝔁𝔁
Friday	𝔁𝔁𝔁𝔁𝔁𝔁𝔁
Saturday	𝔁𝔁𝔁𝔁𝔁𝔁𝔁𝔁

Key: 𝔁 represents 8 people

....

(3)

(Total for Question 1 is 4 marks)

2 Students at The Manor School are assessed in maths.
Each student is awarded a Level.
The table shows this information.

a Complete the table. (2)

b How many students achieved Level 5?

		Level achieved			
Year group	**4**	**5**	**6**	**7**	**8**
7	43	116	28	0	0
8	31	72	63	22	0
9	14	31	85	32	18
Total					

....

(1)

c How many Year 8 students achieved Level 6 or higher?

....

(2)

(Total for Question 2 is 5 marks)

3 Here are some events.

A A coin is spun and shows Heads.

B An ordinary dice is rolled and lands on 5.

C A red card is selected at random from five red cards and one black card.

Show the probability of each event by writing A, B and C in their correct positions on this probability scale.

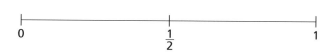

(Total for Question 3 is 3 marks)

4 Mike is an office manager. He wants to know how many letters are sent from his office each day. Design a suitable data collection sheet. Allow for up to 30 letters each day.

(Total for Question 4 is 3 marks)

5 In a school talent contest, each member of the audience voted for their favourite performer. The bar chart shows the results.

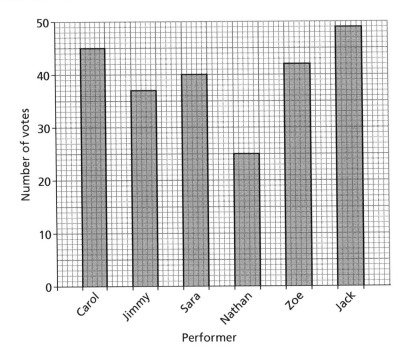

a Who had the highest number of votes?

....
(1)

b How many people voted for Nathan?

....
(1)

c How many more votes did Carol get than Jimmy?

....
(1)

d How many people voted altogether?

....
(1)

(Total for Question 5 is 4 marks)

6 Here are the amounts of weight, in kilograms, lost by 10 dieters in two weeks.

1   5   3   1   2   0   1   4   3   2

   a Write down the mode.

.... kg

(1)

   b Find the median.

.... kg

(1)

   c Work out the mean.

.... kg

(2)

   d Work out the range.

.... kg

(1)

(Total for Question 6 is 5 marks)

7 James has 20 red cards and 10 black cards.
He takes one of the cards at random.
Find the probability that the card will be a red card.

....

(Total for Question 7 is 3 marks)

8 A new housing development has three types of property:
   D detached
   S semi-detached
   T terraced

Each type of property may have one of the following:
   G garage
   P car port
   C conservatory

List all of the possible combinations. The first two are done for you.

DG   DP   ....

....

(Total for Question 8 is 2 marks)

9 In a survey students were asked about their favourite pets.
The dual bar chart shows the results.

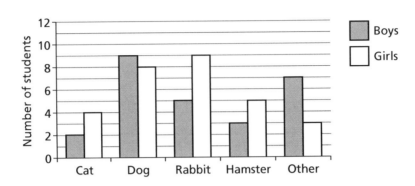

a Which pet did more boys choose than any other?

....
(1)

b How many girls chose hamster?

....
(1)

c How many more girls than boys chose cat?

....
(1)

d Which pet was chosen by the greatest total number of students?

....
(1)

(Total for Question 9 is 4 marks)

10 A small hotel has three floors.
There are three types of room available on each floor.
The two-way table gives information about where the guests stayed one night.

		Type of room		
		Single	**Twin**	**Double**	**Total**
Floor	**3**		8	12	23
	2	2	6		
	1	1		16	
	Total		22		66

a Complete the two-way table.   (2)

A guest is chosen at random.

b Find the probability that the guest has a double room on the second floor.

....

(2)

(Total for Question 10 is 4 marks)

---

11 Vicky has 45 apps on her mobile phone.
The table shows how many she has in each category.

Productivity	Entertainment	Health and fitness	Games
8	10	5	22

Draw a pie chart for this information.

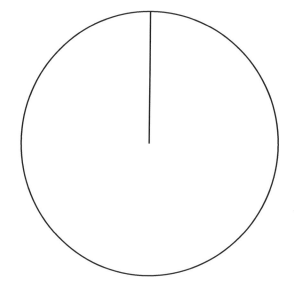

(Total for Question 11 is 4 marks)

12 Jess recorded the lengths of 30 Timber Rattlesnakes.
Here are her results in metres.

1.36	1.27	0.96	1.21	1.05	1.16	1.04	0.98	1.24	1.11
1.18	1.02	1.33	1.24	1.18	1.30	1.12	1.26	0.92	1.00
1.24	1.16	1.12	1.39	1.10	1.27	1.06	1.09	1.28	1.25

Complete the grouped frequency table for the data.

Length (l m)	Tally	Frequency
$0.90 < l \leq 1.00$		
$1.00 < l \leq 1.10$		
$1.10 < l \leq 1.20$		
$1.20 < l \leq 1.30$		
$1.30 < l \leq 1.40$		

(Total for Question 12 is 3 marks)

13 The table shows information about the numbers of children in 30 households.

Number of children	Frequency
0	0
1	10
2	12
3	6
4	2

Work out the total number of children in these households.

(Total for Question 13 is 3 marks)

14 Toby did a survey to find out which type of device people used most often to access the internet. The pie chart shows this information.

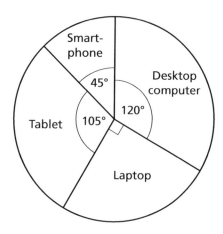

a Which device did most people choose?

....

(1)

Three people chose Smartphone.

b How many chose Tablet?

....

(1)

c How many people were included in the survey?

....

(1)

(Total for Question 14 is 3 marks)

TOTAL FOR TEST IS 50 MARKS

Geometry test

Time: 60 minutes

Edexcel publishes Sample Assessment Material on its website. This Geometry test has been written to help you practise what you have learned and may not be representative of a real exam paper.

1 Here is a line, *AB*.

A _____ B

    a Measure its length.

.... cm

(1)

    b Mark the midpoint of the line with a dot and label it *M*.

(1)

(Total for Question 1 is 2 marks)

2 a Alex is facing South.
He turns 90° clockwise.
In which direction is he now facing?

....

(1)

    b Beth is facing North.
She turns anticlockwise to face East.
What angle does she turn through?

....

(1)

(Total for Question 2 is 2 marks)

3 a Change 5.3 kg to grams.

.... grams

(1)

    b Change 814 cm to metres.

.... metres

(1)

(Total for Question 3 is 2 marks)

4 a Here is a scale measuring in centimetres.

30 cm 40 cm 50 cm 60 cm 70 cm 80 cm 90 c

Mark, with an arrow, a reading of 62 cm.

(1)

b Write down the weight of the tomatoes shown by this scale.

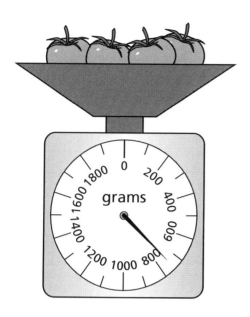

.... grams

(1)

(Total for Question 4 is 2 marks)

---

5 Here is a diagram of a circle, centre O.

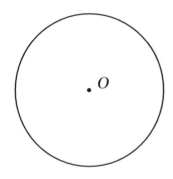

a Draw a diameter. Label it AB. (1)

b Draw a chord. Label it CD. (1)

c Mark a point P on the circumference.
Draw a tangent at P. (1)

(Total for Question 5 is 3 marks)

6 Write acute, obtuse or reflex next to these angles.

a

....
(1)

b

....
(1)

c

....
(1)

(Total for Question 6 is 3 marks)

---

7 Here are some metric units.

centimetres   metres   kilometres   grams   kilograms   millilitres   litres

Select an appropriate unit for measuring these.

a The height of a door

....
(1)

b The mass of a car

....
(1)

c The capacity of an oil drum

....
(1)

(Total for Question 7 is 3 marks)

8 Work out the sizes of angles a and b.

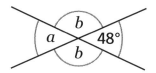

angle $a =$ .... °

(1)

angle $b =$ .... °

(1)

(Total for Question 8 is 2 marks)

9 The grids show a shape and its enlargement (shaded).
What is the scale factor of the enlargement in each case?

a

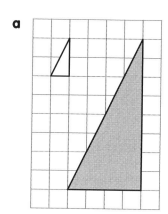

Scale factor = ...

(1)

b

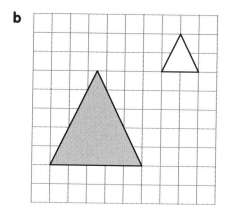

Scale factor = ...

(1)

(Total for Question 9 is 2 marks)

10 Work out the size of angle *c*.

angle *c* = ... °

(Total for Question 10 is 2 marks)

11 For each of these quadrilaterals, put a tick to show which statements are true.

	Rectangle	Parallelogram	Rhombus
All sides equal			
Opposite angles equal			
2 lines of symmetry			

(Total for Question 11 is 3 marks)

12 Work out the size of angle *d*.

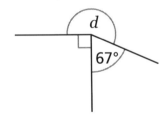

angle *d* = ... °

(Total for Question 12 is 2 marks)

13 a Shade 2 more squares so that this grid has 1 line of symmetry.

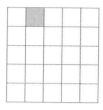

(2)

b Shade 3 more squares so that this grid has 2 lines of symmetry.

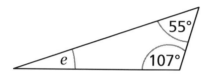

(2)

(Total for Question 13 is 4 marks)

14 Work out the size of angle *e*.

angle *e* = ... °

(Total for Question 14 is 2 marks)

15 Reflect these shapes in the mirror line (shown dashed).

a

(1)

b

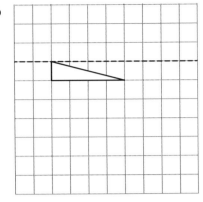

(1)

c

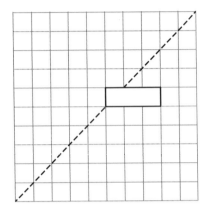

(1)

(Total for Question 15 is 3 marks)

16 a Make an accurate drawing of this triangle.

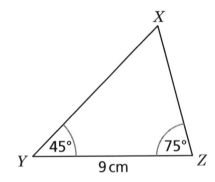

(2)

b Measure the length of XY.

$XY =$ .... cm

(1)

(Total for Question 16 is 3 marks)

17 Work out the size of angle f.

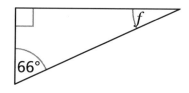

angle $f =$ .... °

(Total for Question 17 is 2 marks)

18 a Work out the area of this rectangle.

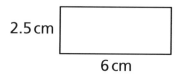

2.5 cm

6 cm

Area = ... cm²

(2)

b Work out the perimeter of this shape.

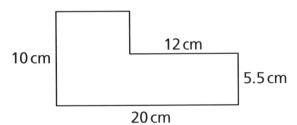

Perimeter = ... cm

(2)

(Total for Question 18 is 4 marks)

19 Work out the size of angle g.

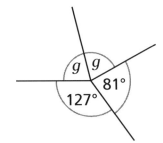

angle g = ... °

(Total for Question 19 is 2 marks)

20 Work out the volume of this cuboid.

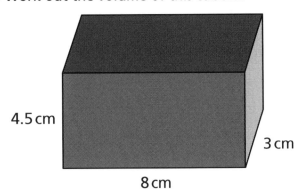

Volume = ... cm^3

(Total for Question 20 is 2 marks)

TOTAL FOR TEST IS 50 MARKS

Answers

1 Data

1.1 Types of data

1 The colour of a car → Discrete
   The weight of an apple → Continuous

2 a Discrete **b** Continuous **c** Continuous
   d Discrete **e** Discrete **f** Continuous

1.2 Data collection

1 a

Number rolled	Tally	Frequency
1	ЖЖ ЖЖ	10
2	ЖЖ IIII	9
3	ЖЖ ЖЖ I	11
4	ЖЖ IIII	9
5	ЖЖ ЖЖ I	11
6	ЖЖ ЖЖ	10

b Total = 60

c The expected total is 60 since the dice was rolled 60 times.

2 a

Guess	Tally	Frequency
301–330	III	3
331–360	III	3
361–390	ЖЖ ЖЖ ЖЖ I	16
391–420	ЖЖ III	8
421–450	ЖЖ ЖЖ I	11
451–480	ЖЖ	5

b 46

3

Height (h m)	Tally	Frequency
$1.60 < h \leqslant 1.65$	III	3
$1.65 < h \leqslant 1.70$	IIII	4
$1.70 < h \leqslant 1.75$	ЖЖ	5
$1.75 < h \leqslant 1.80$	ЖЖ	5
$1.80 < h \leqslant 1.85$	ЖЖ I	6
$1.85 < h \leqslant 1.90$	IIII	4
$1.90 < h \leqslant 1.95$	III	3

4 a

Number spun	Tally	Frequency
1	ЖЖ II	7
2	ЖЖ ЖЖ	10
3	ЖЖ II	7
4	ЖЖ III	8
5	ЖЖ III	8

b 2

5

Number of texts	Tally	Frequency
11–40	ЖЖ II	7
41–80	ЖЖ I	6
81–110	ЖЖ III	8
111–150	IIII	4
151–200	ЖЖ	5

b It is sensible to group the data because the values are so varied.

c 5

6

Number of goals	Tally	Frequency
0	ЖЖ II	7
1	ЖЖ ЖЖ	10
2	ЖЖ ЖЖ II	12
3	ЖЖ	5
4	IIII	4
5	I	1
6	I	1

Don't forget!

* specific
* numerical, range
* grouped, class

Unit test

1 a

Vehicle	Tally	Frequency
Bus	II	2
Car	ЖЖ ЖЖ ЖЖ ЖЖ ЖЖ II	27
Motorbike	II	2
Truck	ЖЖ IIII	9
Van	ЖЖ ЖЖ	10

b Categorical

2

Length (l cm)	Tally	Frequency
$12.0 < l \leqslant 13.0$	III	3
$13.0 < l \leqslant 14.0$	II	2
$14.0 < l \leqslant 15.0$	IIII	4
$15.0 < l \leqslant 16.0$	II	2
$16.0 < l \leqslant 17.0$	ЖЖ	5
$17.0 < l \leqslant 18.0$	IIII	4
$18.0 < l \leqslant 19.0$	ЖЖ	5
$19.0 < l \leqslant 20.0$	ЖЖ	5

2 Displaying data

2.1 Pictograms

1 Pictogram showing people attending the school production

Monday	☺☺☺☺☺☺☺☺☺☺☺☺
Tuesday	☺☺☺☺☺☺☺☺☺
Wednesday	☺☺☺☺☺☺☺☺☺☺
Thursday	☺☺☺☺☺☺☺☺
Friday	☺☺☺☺☺☺☺☺☺☺☺☺
Saturday	☺☺☺☺☺☺☺☺☺☺☺☺☺☺☺

Key: ☺ represents 10 people

2 b  **c** **d**

	Time	Number of customers
3	7 pm	⊕⊕⊕⊕⊕
	8 pm	⊕⊕⊕⊕⊕⊕⊕⊕⊕◖
	9 pm	⊕⊕⊕⊕⊕⊕⊕⊕⊕⊕⊕◖
	10 pm	⊕⊕⊕⊕⊕⊕⊕⊕
	11 pm	⊕⊕⊕⊕⊕⊕◖

Key: ⊕ represents 4 customers

	Day	Number of letters received
4	Monday	☐☐☐☐☐☐☐☐☐
	Tuesday	☐☐☐☐☐
	Wednesday	☐☐☐☐☐☐☐☐
	Thursday	☐☐☐☐☐
	Friday	☐☐☐☐☐☐

Key: ☐ represents 2 letters

2.2 Bar charts

1 Bar chart to show the colours of sweets in a jar

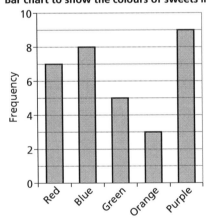

2 a

Colour	Blue	Green	Yellow	Red
Frequency	8	3	9	14

b

Item	Rent	Food	Clothes	Entertainment	Other
Percentage	28	24	16	21	11

3

Programme	Comedy	Drama	Soap	Film	Documentary
Frequency	11	6	10	9	4

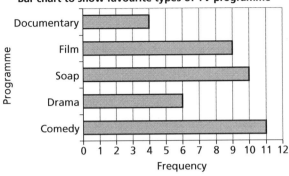

Bar chart to show favourite types of TV programme

4

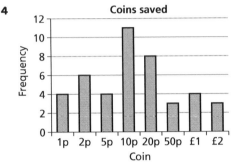

Coins saved

2.3 Line graphs

1

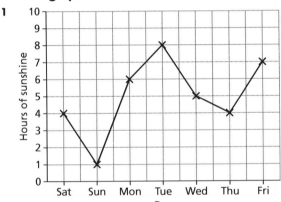

2

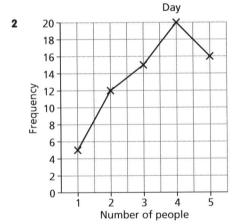

3

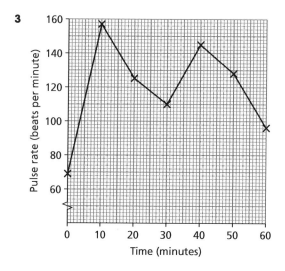

4

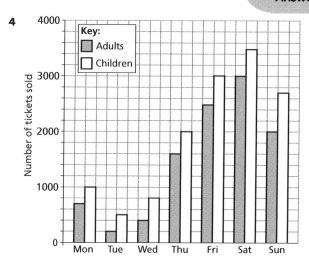

2.4 Dual bar charts

1

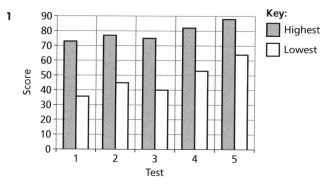

2

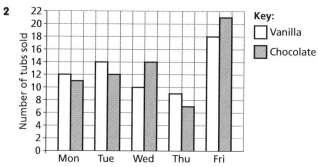

3

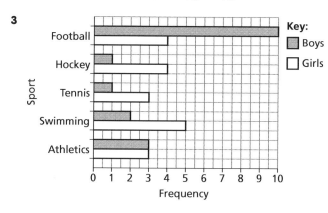

2.5 Two-way tables

1

	Art	Biol	Chem	Econ	Eng	Geog	Hist	Maths	Phys	Total
Boys	1	3	1	2	3	3	3	4	2	22
Girls	3	4	1	1	3	1	3	4	0	20
Total	4	7	2	3	6	4	6	8	2	42

2

	French	German	Spanish	Total
Boys	7	6	5	18
Girls	8	7	9	24
Total	15	13	14	42

3

	Cat	Dog	Fish	Hamster	Mouse	Rabbit	Rat	Total
Boys	3	7	3	2	2	2	2	21
Girls	4	5	1	3	4	4	0	21
Total	7	12	4	5	6	6	2	42

4

	Sixth form	College	Work	Total
Boys	24	23	11	58
Girls	28	29	16	73
Total	52	52	27	131

5

	Did homework	Did not do homework	Total
Did bring equipment	24	2	26
Did not bring equipment	1	4	5
Total	25	6	31

2.6 Pie charts

1

Method of travel	Number of students	Angle of pie chart
Bus	14	12° × 14 = 168°
Car	5	12° × 5 = 60°
Cycle	3	12° × 3 = 36°
Walk	8	12° × 8 = 96°
Total	30	12° × 30 = 360°

2

Fuel type	Number of vehicles	Angle of pie chart
Petrol	72	216°
Diesel	31	93°
Dual fuel	2	6°
Hybrid	15	45°
Total	120	360°

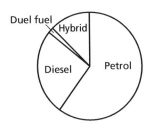

3

Waste type	Number of containers	Angle of pie chart
General waste	30	150°
Metal	8	40°
Organic	16	80°
Paper	6	30°
Wood	12	60°
Total	72	360°

4

Activity	Number of hours	Angle of pie chart
Sleeping	9	135°
Working	7	105°
Dining	2	30°
Relaxing	6	90°
Total	24	360°

Don't forget!

* key
* gap
* discrete
* comparisons
* total, columns
* small

Unit test

1 a

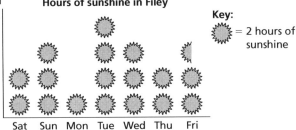

Hours of sunshine in Filey

b

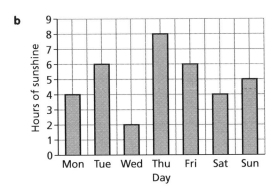

2

	Junior	Adult	Senior	Total
Standard	36	16	5	57
Full	5	18	10	33
Premium	0	25	0	25
Total	41	59	15	115

3

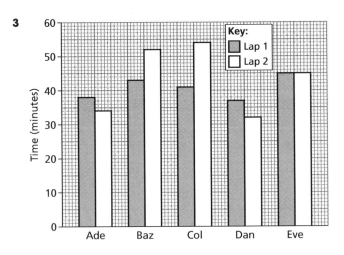

4

Book category	Number of books	Angle of pie chart
Comedy	12	96°
Thriller	7	56°
Mystery	9	72°
Romance	17	136°

3 Calculating with data

3.1 Averages and range

1 white
2 1
3 4
4 3
5 48
6 blue
7 9, 11, 12, 15, 16, 16, 17, 18, 18. The median is 16
8 7, 7, 8, 8, 9, 9, 10, 10.
  The numbers in the middle are 8 and 9. The median is 8.5
9 Median height = 180 cm
10 Median number of eggs = 8
11 Median number of press-ups = 21
12 Median cooking time = 15.5 minutes
13 Total = 3 + 0 + 2 + 1 + 5 + 2 + 1 + 2 = 16.
  Mean = 16 ÷ 8 = 2
14 Total = 3 + 5 + 1 + 7 + 4 + 10 = 30. Mean = 30 ÷ 6 = 5
15 Mean = 5
16 Mean = 12
17 Mean = 24
18 Mean = 3.6
19 Largest salary = £1105. Smallest salary = £824.
  Range = £1105 − £824 = £281
20 Range = 59
21 Range = 3.5 m
22 Range = 56 minutes
23 Range = 57 seconds
24 Range = 12°C

Don't forget!

** average
* most often
* order
* total of all the data values, number of values
* range

Unit test

1 a Mode = 1
  b Median = 2
  c Mean = 2.27 to 3 sf
2 a Mean = 59
  b Median = 58
  c Range = 24
3 a Mode = 23 seconds
  b Median = 23 seconds
  c Mean = 23 seconds

4 Interpreting data

4.1 Read and interpret data from tables

1 a 124
  b Year 9
  c 124 + 116 = 240
2 a 1303
  b 1320 − 3 = 1317
3 a 547
  b 1883
  c 1254
4 a 113 miles
  b York
  c 122 + 113 = 235 miles
5 a −2°C
  b Cardiff
  c 4°C
6 a 0653
  b 22 minutes
7 a 3
  b 16
  c Jinty
  d 4

8 a 64 miles
  b Worcester
  c Sheffield
  d 298 miles
  e 49 miles
9 a 4
  b 2000
  c 4
  d 46
10 a Rum and raisin
  b 142
  c July
  d 558 litres
  e 139 litres

4.2 Interpret charts and graphs

1 a 45
  b Saturday
  c 15
  d 235
2 a 250 000 tonnes
  b 200 000 tonnes
  c 75 000 tonnes
3 a 3 million
  b British Museum
  c 500 000 or 0.5 million
  d 24 500 000 or 24.5 million
4 a Rottweiler
  b 7 years
  c 8.5 years
  d Bulldog
5 a Toyota Prius
  b Ferrari 458
6 a £60 000 000 or £60 million
  b 2009
  c 2011
7 a 4
  b Level 6
  c 20
  d 31
8 a 5
  b 8
  c 7
  d 30
  e Answer should include the fact that pupils switched from choosing numbers in the middle of the range to numbers at the ends.
9 a 4
  b Answer should mention that nothing is known about the number of customers at times between the plotted points. There may have been times when there were no customers.
10 a 800 000
  b 300 000
  c 1980
  d 40 (accept any answer between 36 and 46)
11 a Diesel
  b 2010 and 2011
  c 2009
12 a 23
  b 5
  c 2
  d Week 4
  e 9
13 a Learn to drive (largest angle)
  b Learn self-defence (same angle as 'Learn to dance')
  c 9 × 6 = 54
14 a 42
  b 21
  c 7
15 a £90
  b £45
  c £60

4.3 Find totals and modes from frequency tables or diagrams

1 a 25

b

Goals scored (G)	Frequency (F)	G × F
1	5	5
2	8	16
3	6	18
4	4	16
5	2	10
Totals	25	65

Total number of goals scored = 65

2

Number of people (P)	Frequency (F)	P × F
1	32	32
2	21	42
3	12	36
4	8	32
5	3	15
	76	157

a There were 76 cars in the survey
b The total number of people was 157

3 a 58
b 1663

4 3

5 Pop

6 a 20
b 48
c 2

7 a 94
b 240
c 2

Don't forget!

* title
* pattern
* total
* highest

Unit test

1 a Missing values are: Germany 11 Gold;
Republic of Korea 7 Bronze; Great Britain 65 total;
United States of America 104 total.
b The order is decided on the number of Gold medals won.
Great Britain won more Gold medals than the Russian
Federation.
c 16

2 a 235 miles
b Bristol
c 293 miles

3 a 600
b

Monday	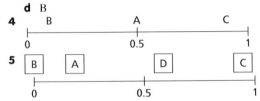
Tuesday	
Wednesday	
Thursday	
Friday	
Saturday	

Key:
○ represents 400 people

c 300

4 a Cycling
b Taekwondo
c 20

5 a Missing value: 3
b 3
c 4

5 Probability

5.1 Use and interpret a probability scale

1 b Unlikely
c Certain
d Impossible

2 a Likely
b Evens
c Unlikely

3 a C
b A
c A
d B

4 B ⟷ A ⟷ C on scale 0, 0.5, 1

5 B A D C on scale 0, 0.5, 1

5.2 Write down theoretical and experimental probabilities

1 a $\frac{1}{6}$
b $\frac{3}{6}$ or $\frac{1}{2}$
c $\frac{4}{6}$ or $\frac{2}{3}$
d $\frac{5}{6}$
e 0

2 a $\frac{2}{5}$
b $\frac{3}{5}$
c $\frac{3}{5}$
d $\frac{1}{5}$
e 0

3 a $\frac{8}{20}$ or $\frac{4}{10}$
c $\frac{15}{20}$ or $\frac{3}{4}$
b, d 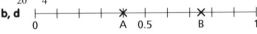 A 0.5 B

4 a $\frac{18}{42}$ or $\frac{3}{7}$
b $\frac{8}{42}$ or $\frac{4}{21}$
c $\frac{14}{42}$ or $\frac{1}{3}$

5 a $\frac{7}{25}$
b $\frac{8}{25}$
c $\frac{5}{25}$ or $\frac{1}{5}$

6 a $\frac{31}{50}$
b $\frac{19}{50}$

7 a $\frac{8}{40}$ or $\frac{1}{5}$
b $\frac{17}{40}$
c $\frac{28}{40}$ or $\frac{7}{10}$

8 a $\frac{1}{20}$
b $\frac{5}{20}$ or $\frac{1}{4}$
c Answer should include: 6 has a much higher relative frequency
than 5; however, 20 is a small number of trials so the results are
unreliable.
d Megan should carry out a lot more trials to make her results
more reliable.

5.3 List outcomes

2 1, 2, 3, 4, 5, 6
3 red, red, yellow, yellow, yellow
4 blue, blue, green, green, green, green
5 red, white, blue
6 1, 2, 3, 4
7 H1, H2, H3, H4, H5, H6; T1, T2, T3, T4, T5, T6
8 HH, HT, TH, TT
9 SB, SR, SW; EB, ER, EW; HB, HR, HW
10 QX, QY, WX, WY, EX, EY, RX, RY
11 (A, 1), (A, 2), (A, 3), (B, 1), (B, 2), (B, 3), (C, 1), (C, 2), (C, 3),

Don't forget!

* outcomes
* 0
* 0, 1
* the event, number of outcomes
* The total number of trials
* increased

Unit test

1
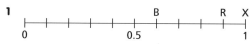

2 a 28
b $\frac{9}{28}$
c $\frac{22}{28}$ or $\frac{11}{14}$
d $\frac{7}{28}$ or $\frac{1}{4}$
e 0
3 SD, SG, SP, FD, FG, FP, ND, NG, NP

6 Measures

6.1 Metric and imperial measures

1 a length: door, ribbon   weight: potatoes, cement
capacity: water, bottle
2 a cm, m   **b** grams, kg   **c** ml/grams   **d** km
e litres/kg   **f** grams/cm
3 a 250 ml   **b** 180 g   **c** 16 m

6.2 Convert between metric units

1 a 3 cm   **b** 7.5 cm   **c** 60 mm   **d** 25 mm
e 12.4 cm   **f** 257 mm   **g** 3 m   **h** 3.5 m
i 3.54 m   **j** 400 cm   **k** 450 cm   **l** 458 cm
2 a 4 km   **b** 4.5 kg   **c** 4.567 litres   **d** 3000 m
e 5600 g   **f** 2345 ml   **g** 7200 m   **h** 6 kg
i 7.2 litres
3 a 250 ml, 0.4 litres, 1800 ml, 2.5 litres
b 250 cm = 2.5 m
25 cm = 250 mm
0.25 km = 250 m
1250 m = 1.25 km

6.3 Add and subtract units of measure

1 a 69 mm   **b** 77 mm   **c** 7 m 65 cm
d 8 m 10 cm   **e** 243 cm   **f** 552 cm
2 a 854 cm   **b** 255 g   **c** 100 mm
d 1.3 km   **e** 60 mm   **f** 2 litres 800 ml
g 1 m 70 cm   **h** 3 m 40 cm
3 a 5.7 m or 570 cm   **b** 2.15 km   **c** 600 ml or 0.6 litre
d 4.7 m or 470 cm

6.4 Read scales

1 a 2   **b** 5   **c** 20   **d** 100 ml   **e** 250 g
2 a 1 litre 400 ml   **b** 800 g
c 1 m 75 cm   **d** 4.5 cm = 45 mm
3 a    **b**

c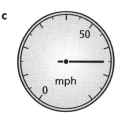

d

e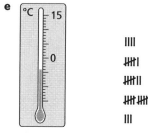

4 a 650 g; 1 litre 300 ml
b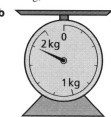

c 77 or 78 cm

6.5 Draw and measure lines and angles

Diagrams in the following section are not drawn to scale. Please check you have drawn the lines and angles in your own constructions accurately according to the instructions in the questions.

1 a 4 cm   **b** 6 cm   **c** 7 cm
2 a A————————————B
b C————————————D
c E————————————F
3 a 4 cm, 3 cm, 5 cm; AC
4 L————————M————

Don't forget!

* gram → weight; litre → capacity; kilogram → weight;
mililitre → capacity; kilometre → distance/length;
cm → distance/length; milimetre → distance/length;
litre → capacity
* multiply; divide

Unit test

1 a 5 cm   **b** 3.5 cm   **c** 4 cm
2 a 4 litres   **b** 95 mm
3 6 m 55 cm
4 a cm or metres   **b** litres   **c** grams
5 a 1 m 25 cm   **b** 1 kg 200 g   **c** 750 ml

7 Angles

7.1 Angles and turning

1 a $\frac{1}{4}$ turn **b** $\frac{1}{2}$ turn **c** $\frac{3}{4}$ turn **d** full turn

2 a South **b** West **c** West **d** East

3 a $\frac{1}{2}$ turn clockwise **b** $\frac{3}{4}$ turn anticlockwise

 c $\frac{1}{4}$ turn clockwise **d** $\frac{3}{4}$ turn clockwise

4 a anticlockwise **b** $\frac{1}{4}$ turn **c** $\frac{3}{4}$ turn clockwise

7.2 Naming, measuring and drawing angles

1 a obtuse **b** reflex **c** acute **d** reflex

2 a reflex **b** obtuse **c** obtuse **d** acute

3 angle *DAB* is an acute angle

4 a angle *QPR* is an obtuse angle

 b angle *YZW* is an acute angle

 c angle *KMN* is a right angle

 d angle *STU* is a reflex angle

5 a There are 4 acute angles, they are angle *ABC*, angle *CDE*, angle *DEF*, angle *FGA*

 b There is 1 obtuse angle, angle *BAG*

 c There are 2 reflex angles, they are angle *BCD*, angle *EFG*

6 a 60° **b** 150°

7 a 75° **b** 125° **c** 118° **d** 23°

 e 136° **f** 109°

8 *A* 110°, *B* 45°, *C* 25°

9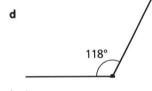

10 a

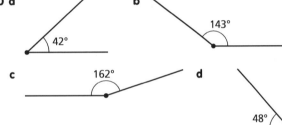

11

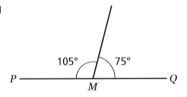

Don't forget!

* acute, obtuse, reflex

Unit Test

1 a acute **b** reflex **c** obtuse **d** reflex

2 a angle *DGC*, obtuse **b** angle *PFT*, reflex

 c angle *KHE*, acute **d** angle *YWM*, right angle

3 a 35° **b** 138°

 Angles drawn to ± 1° accuracy

4 a **b**

Angles drawn to ± 1° accuracy

8 Angle Calculations

8.1 Angles on a straight line and angles around a point

1 $a = 130°$

2 $b = 44°$

3 $c = 63°$

4 $d = 102°$

5 $e = 32°$

6 $f = 38°$

7 $g = 85°$

8 $h = 46°, j = 134°$

9 $k = 146°$

10 $m = 63°$

11 $n = 35°$

8.2 Triangle properties

1 B because A and C are both equilateral, B is isosceles

2 B because B is isosceles, A and C are right angled but not isosceles

3 a **b** **c**

4 a Equilateral ... triangle *ADF* **b** Right angled ... triangle *DEF*

 c Scalene ... triangle *ABC* **d** Isosceles ... triangle *ACD*

5 $m = 71°$

6 $n = 128°$

7 $p = 56°$

8 $t = 70°$

Don't forget!

* 180°, 180°

* 360°, 360°

Unit test

1 $g = 113°$ **2** $h = 72°$ **3** $p = 48°$

4 $w = 115°$ **5** $x = 53°, y = 127°$ **6** $z = 62°$

7 B is equilateral, the others are isosceles

8 $k = 47°$ **9** $y = 39°$ **10** $a = 18°$

11 $m = 42°$ **12** $c = 110°$

9 Constructing Triangles

9.1 Accurate drawings

All triangles drawn here are sketches of the correct answers; they are not full size, nor accurate but the students' diagrams must be drawn full size, with angles correct to ± 1°

1

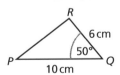

$BC = 6.9$ cm

2

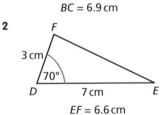

$EF = 6.6$ cm

3

$PR = 7.7$ cm

4

$JL = 6.6$ cm $KL = 4.9$ cm

5

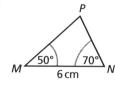

$MP = 6.5\,cm$ $NP = 5.3\,cm$

6

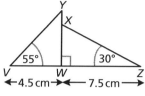

$VY = 7.8\,cm$ $XZ = 8.7\,cm$ $XY = 2.1\,cm$

7

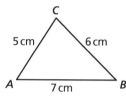

angle $A = 57°$

8

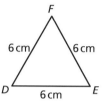

All angles are 60°
This is an equilateral
triangle

9

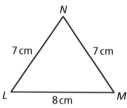

angle $L = 55°$ angle $M = 55°$

This is an isosceles triangle

10

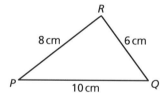

angle $RPQ = 37°$ angle $PQR = 53°$ angle $PRQ = 90°$

Unit test

1

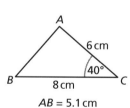

$AB = 5.1\,cm$

2

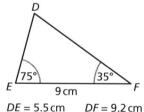

$DE = 5.5\,cm$ $DF = 9.2\,cm$
                    or $9.3\,cm$
This is a scalene triangle

3

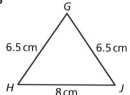

angle $H = 52°$ angle $J = 52°$
This is an isosceles triangle

4

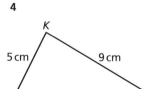

angle $K = 86°$ angle $L = 64°$
angle $M = 30°$

10 Transformations

10.1 Reflection

1 **a** **b** **c**

2 **a**  **b** **c**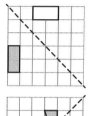

3 **a** 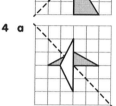 **b**

4 **a** 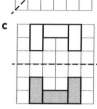 **b**

10.2 Enlargement

1 **a** Scale factor 3   **b** 8 squares long; $8 \div 4 = 2$;
                              Scale factor 2
2 **a** Scale factor 5   **b** Scale factor 3
3 **a** Scale factor 2   **b** Scale factor 3
    c Scale factor $1\frac{1}{2}$ **d** Scale factor $2\frac{1}{2}$

10.3 Congruent shapes and similar shapes

1 The shapes **are** congruent.
2 **a** Yes   **b** No   **c** Yes   **d** Yes
    e Yes   **f** No
3 **a** They are **not** congruent   **b** The shapes **are** similar
4 $BC = 1.5\,cm$, $QR = 4.5\,cm$
    $PQ = 2 \times AB$, $QR = 3 \times BC$
    The scale factors are **different** so the shapes are **not** similar.
5 **a** No
    b Yes, scale factor $= 2$
    c Yes, scale factor $= \frac{1}{2}$
    d Yes, scale factor $= \frac{1}{2}$
    e Yes, scale factor $= \frac{1}{2}$
    f Yes, scale factor $= 2\frac{1}{2}$

Don't forget!

* a line of symmetry
* sizes
* identical
* angles
* same

Unit test

1

2 a Scale factor = 2 **b** Scale factor = $1\frac{1}{2}$ **c** Scale factor = $2\frac{1}{2}$

3 a Scale factor = 2 **b** Scale factor = $3\frac{1}{2}$ **c** Scale factor = $1\frac{1}{2}$

4 a No **b** Yes **c** Yes
d Yes **e** Yes **f** No

5 a Yes, scale factor = $1\frac{1}{2}$ **b** Yes, scale factor = 2
c No **d** Yes, scale factor = 2
e Yes, scale factor = 2 **f** Yes, scale factor = 3

11 Circle and quadrilateral definitions

11.1 Circles

1 Circle, diameter 4 cm with centre and radius marked, as in text.

2 a 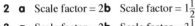 **b**

3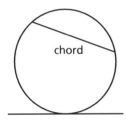

4

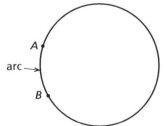

5

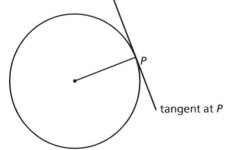

11.2 Quadrilaterals

1 Rhombus
2 a Rectangle or Square **b** Trapezium
c Parallelogram or Rhombus or Rectangle or Square
d Parallelogram or Rhombus **e** Kite
3 a

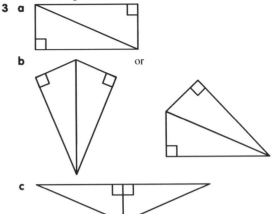

b or

c

Don't forget!

* Centre
* radius
* edge
* touches
* chord
* circumference
* 4

Unit test

1 *AC* is a diameter, *OB* is a radius, *AD* is a chord, *PQ* is a tangent, from *C* to *D* is an arc
2

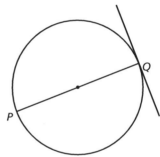

3 Square or Rhombus
4

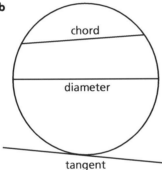

or

12 Perimeter, area and volume

12.1 Perimeter of rectangles

1 a $5\,\text{cm} + 7\,\text{cm} + 5\,\text{cm} + 7\,\text{cm} = 24\,\text{cm}$
b $12\,\text{m} + 3\,\text{m} + 12\,\text{m} + 3\,\text{m} = 30\,\text{m}$
c $26\,\text{cm} + 14\,\text{cm} + 26\,\text{cm} + 14\,\text{cm} = 80\,\text{cm}$
2 a 178 cm **b** 33.6 m
3 a 20 cm **b** 29.6 cm
4 a 80 m **b** 170 cm
5 a 12 cm **b** 6 cm
6 $42 + 20 = 62\,\text{cm}$; $23 - 12 = 11\,\text{cm}$;
$62 + 23 + 20 + 11 + 42 + 12 = 170\,\text{cm}$
7 116 cm
8 60 cm

12.2 Area of rectangles

1 a $6\,\text{cm}^2$ **b** $4 \times 6 = 24\,\text{m}^2$ **c** $5 \times 3 = 15\,\text{mm}^2$
2 a $18\,\text{cm}^2$ **b** $20\,\text{m}^2$ **c** $6\,\text{mm}^2$
3 a $27.28\,\text{m}^2$ **b** $45\,\text{cm}^2$

12.3 Volume of cuboids

1 a $8\,cm^3$ b $12\,cm^3$ c $3 \times 2 \times 4 = 24\,cm^3$
2 a $30\,cm^3$ b $240\,cm^3$ c $3300\,cm^3$
3 a $86\,625\,mm^3$ b $100.44\,m^3$ c $98.716\,cm^3$
4 a $l = w = h = 1\,cm, V = 1\,cm^3$;
   $l = w = h = 2\,cm, V = 8\,cm^3$;
   $l = w = h = 3\,cm, V = 27\ cm^3$;
   $l = w = h = 4\,cm, V = 64\,cm^3$;
   $l = w = h = 5\,cm, V = 125\ cm^3$;
   $l = w = h = 6\,cm, V = 216\ cm^3$;
   $l = w = h = 7\,cm, V = 343\ cm^3$;
   $l = w = h = 8\,cm, V = 512\ cm^3$;
   $l = w = h = 9\,cm, V = 729\ cm^3$;
   $l = w = h = 10\,cm, V = 1000\ cm^3$
   b $5\,cm$
   c $10\,m$

Don't forget!

* add width and length, then double the total
* multiply the width by the length
* multiply length $\times$ width $\times$ height
* all its edges are the same length

Unit test

1 a $16\,cm$ b $12\,cm^2$
2 a $22\,cm$ b $18\,cm^2$
3 $44\,cm$
4 $90\,cm^3$
5 $64\,m^3$
6 $108\,cm^3$

Statistics Test

1 a Show for Wednesday on the pictogram.
   b 264
2 a Totals are: Level 4: 88; Level 5: 219; Level 6: 176; Level 7: 54; Level 8: 18
   b 219
   c 85
3

```
   B        A        C
├──────────┼──────────┤
0         ½          1
```

4 Data should be grouped, but different intervals may be used.

Number of letters	Tally	Frequency
0−5		
6−10		
11−15		
16−20		
21−25		
26−30		

5 a Jack
   b 25
   c 8
   d 238
6 a $1\,kg$
   b $2.5\,kg$
   c $2.2\,kg$
   d $5\,kg$
7 $\frac{20}{30}$ or $\frac{2}{3}$
8 DG, DP, DC, SG, SP, SC, TG, TP, TC
9 a dog
   b 5
   c 2
   d dog

10 a

		Type of room			
		Single	Twin	Double	Total
Floor	3	3	8	12	23
	2	2	6	10	18
	1	1	8	16	25
	Total	6	22	38	66

   b $\frac{10}{66}$ or $\frac{5}{33}$

11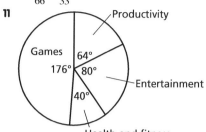

12

Length (l m)	Tally	Frequency
$0.90 < l \leqslant 1.00$		4
$1.00 < l \leqslant 1.10$		6
$1.10 < l \leqslant 1.20$		7
$1.20 < l \leqslant 1.30$		10
$1.30 < l \leqslant 1.40$		3

13 Missing number in table is 2. Total = 60
14 a Desktop computer
    b 7
    c 24

Geometry test

1 a $7\,cm$
   b

```
              M
A ────────────┼──────────── B
         (allow± 1 mm)
```

2 a West b 270°
3 a 5300 grams b 8.14 metres
4 a Arrow at one gradation after 60
   b 750 grams (allow ± 5 grams)
5 a Line through centre, full diameter drawn, labelled AB
   b Any chord drawn (not a diameter), labelled CD
   c Line touching circle at any point marked P.
6 a Obtuse b Acute c Reflex
7 a metres b kilograms c litres
8 angle $a = 48°$ angle $b = 132°$
9 a Scale factor = 4 b Scale factor = $2\frac{1}{2}$
10 angle $c = 29°$
11

	Rectangle	Parallelogram	Rhombus
All sides equal			√
Opposite angles equal	√	√	√
2 lines of symmetry	√		√

12 angle $d = 203°$

13 a Any of these six possible answers

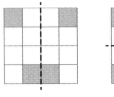

b

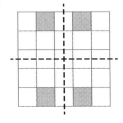

14 angle $e = 18°$

15 a 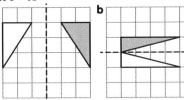 **b** **c**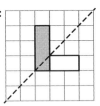

16 a This triangle drawn accurately and the correct size.

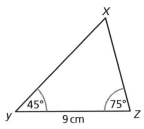

b $XY = 10$ cm (10.038...)

17 angle $f = 24°$

18 a 15 cm^2 **b** 60 cm

19 angle $g = 76°$

20 108 cm^3

Printed in Great Britain
by Amazon